The Nature of

Florida's Neighborhoods

Including

Bats, Scrub jays, Lizards and Wildflowers

by

Atlantic Press

PO Box 510366

Melbourne Beach, Florida 32951

Other books by Cathie Katz

The Nature of Florida's Beaches

The Nature of Florida's Waterways

The Nature of Florida's
Neighborhoods

Including
Bats, Scrub jays, Lizards and Wildflowers

Written and Illustrated by Cathie Katz

Atlantic Press

PO Box 510366

Melbourne Beach, Florida 32951

Printed on recycled paper

The Nature of Florida's Neighborhoods
Including Bats, Scrub jays, Lizards and Wildflowers

Copyright by Cathie Katz © 1997
Photographs by Jim Angy © 1997
Poems by Patricia Ryan Frazier © 1997

Published by
Atlantic Press
PO Box 510366
Melbourne Beach, Florida 32951

Printed in the United States of America.

The information about edibles in this book is general. Any plant used externally or internally can cause allergic reactions in some people. Neither the publisher nor the author accept responsibility for any harm resulting from mistaken identity or inappropriate use of any wild edible.

Cataloging-in-Publication Data
    Katz, Cathie
    The Nature of Florida's Neighborhoods Including Bats, Scrub jays, Lizards and Wildflowers

    Includes bibliography and index.
    p.   cm.

    1. Natural history - Florida.   2. Coastal Flora - Florida. 3. Birds - Florida. 4. Insects - Florida. 5. Mammals - Florida. 6. Reptiles - Florida. I. Title.

508.759
ISBN Number: 1-888025-09-3   Library of Congress Catalog Card Number: 96-084821

The Nature of Florida's Neighborhoods

Contents

Florida's Neighborhoods

INTRODUCTION

The Nature of Florida's Neighborhoods is the third in a series of books about Florida's habitats. The idea to write about our neighborhood as a habitat came to me last year as I was working in my outside office (a lawn chair and PVC table). When I noticed the mulch under my desk move, I wondered, "What's going on down there?" It turned out to be a mole cricket crawling back to its burrow. And what an incredible cricket it was!

I was convinced that this amazing creature was the product of a designer with a sense of humor. I wanted to know more about this hidden life under my feet.

Right after that experience, a parade of fire ants found the soft spaces between my toes. Why did their stings burn like that?

Then I heard a blue jay scream. And then a flock of crows started yelling. And then the mockingbirds got into the noise-making ceremony. What were they saying?

Did I really see a scorpion?

Then I wondered, where do maggots come from?

Why do mourning doves squeak like rusty hinges when they take off?

And why was that lump of poop on the citrus tree *crawling*?

And what were those spidery beads dangling under the lawn umbrella?

What does a no-see-um look like?

What makes those shiny trails on the sidewalk?

Do bed bugs really live in beds?

The time had come to answer those questions and all the others I had about the wildlife living around me.

Neighbors

My neighborhood is a 30-mile stretch between my job at Cape Canaveral Air Station to my home in Melbourne Beach. Between these two places, I've learned a lot about our neighborhood habitats, especially that Florida is much much more than manatees, alligators and Mickey Mouse.

I've also discovered the value of my human neighbors too. My work neighbors at The Applied Physics Laboratory have listened politely to my long and detailed stories of cockroach poop and lizard eggs for the past few months without complaining. From their attentiveness (or lack of) I learned what's interesting and what's not. (Not everyone is captivated by green maggot development -- can you imagine?)

And I'm very lucky that one of my neighbors is also my office manager. Sue Bradley keeps my books and office in order, as well as keeping my mental and nutritional health in balance. Thanks to Sue, I always have a supply of artist's pens and paper handy, a refrigerator full of my favorite yoghurt, and lots of encouraging murmurs over the phone when I get weary from too many late nights of work.

And without knowing it, the Melbourne Beach Police Officers contributed a lot to these books too. When I walk through my neighborhood each night, I see them patrolling the streets. They always wave at me (the crazy lady with the white baseball cap). They let me feel secure enough to know that I can get lost inside my head and write these words without being distracted with worry. I owe these officers a big debt.

Thanks neighbors.

Sunrise Symphony

God!
Will I ever
get used to it?
Every day, it's the
grand overture, accompanied
by an arrangement of birds tuning up --
fluttering, twittering, testing -- and then blending
in heaven-bound harmonies, building to a
crescendo as the dark curtain of
night rises and surprises me
with the sun ... bringing
a glorious concert of
light and music into
my life again.
Tomorrow,
Encore!

- Patricia Ryan Frazier

This book is for my family: Toby, Jules, Harpo, Peter, Ellen, Lisa, Chris, Aaron and Rosalee.

BACKYARDS

"We have to get back the balance of nature by protecting wild birds and plants. Our own survival is closely linked with it ... We know that the universe was not created haphazardly."
- Sybil Leek, naturalist and author of *Diary of a Witch*

Little Backyard Birds

Anyone who has birdfeeders knows how difficult it is to keep **squirrels** (*Sylvilagus carolinensis*) away. I tried most of the suggestions such as putting the birdfeeder on a 5-foot greased pole, using plastic bafflers, blasting them with a super watergun. But the squirrels always outsmarted me, so I quit doing battle with them. After I gave up, I wondered what took me so long. Accepting their determination is a lot easier than trying to get my way. Now I put out a dish of peanuts for them near the birdfeeders. We're all happier, especially the birds.

Carolina chickadees (*Parus carolinensis*) are easy to identify by their black-capped heads (shown left).

Chickadee couples stay together all year and hang out with other flocks of small birds like warblers and wrens. Males and females look identical, and both sing lots of different tunes. Even though I can't identify their different calls, I know each tune sends a different message. Some calls are just between courting pairs and others may signal their flock to band together. The flocks are tightly structured units. Like other birds, chickadees have both breeding and non-breeding territories. But, unlike other birds, the whole flock is responsible for defending their territory, rather than just the mated pair.

Other little migrants that we see in the winter are **solitary vireos** (*Vireo solitarius*), **chipping sparrows** (*Spizella passerina*) and **palm warblers** (*Dendroica palmarum*).

Faith is the bird that
Feels the light and sings
When the dawn is still dark
(from Calcutta poet, Sir Rabindranath Tagore)

Backyard Birds

From Shirley Hills in "Nature Notes" in The Indian River Audubon Society's Newsletter, *The Limpkin*: "All of the yard birds love the birdbaths. To prevent waiting lines and battles I added a couple more birdbaths. Although all of them are shallow, several birds show marked preferences for a particular birdbath. Location is probably an important factor. Surprisingly, an old dinner plate filled with water is one of the most popular water spots. Tucked in between potted plants on a low bench next to the window, this simple plate is the perfect bathtub for the Carolina wrens who prefer this secret location."

Carolina wrens (*Thryothorus ludovicianus*) are about 5½ inches and are seen in Florida all year. They often nest in cans, buckets, boxes or old shoes.

Shallow containers of water near or under the edge of a bush will attract local and migratory birds, especially if you include dripping water. The sound of a slow drip from a hose attracts them, and the ripples catch their vision.

Birds' songs

Strangely, I became conscious of Florida bird calls when I was in the mountains of Utah; their bird calls were unfamiliar to me and I realized how accustomed I was to hearing my Florida birds. This started me thinking how much we humans rely on chirps and beeps to communicate too. For example, a beep at the ATM means, *You're finished, take your card and leave;* DING DONG at the door lets me know someone wants to see me; at work my computer says BLEEP, *You've made a mistake;* when the telephone rings I know it means *You have a phone call, pick up the receiver.* We say a lot with one or two notes. We say even more with 26 letters (everything actually). Imagine how much information birds can give with their range of notes.

"Migration is a headache that birds get when they fly south for the winter."
-A student's exam answer recorded in *More Anguished English* by Richard Lederer

Painted Buntings

From "The Most Beautiful Bird in the World" by Bob Brown in *The Limpkin*:
"We refer to the bald eagle as majestic. The peregrine falcon is powerful. The scrub jay is certainly wonderful and tame enough to eat out of one's hand ... But only one holds the distinction as being the most beautiful bird in the world ... that bird is our painted bunting ... the nonpareil."

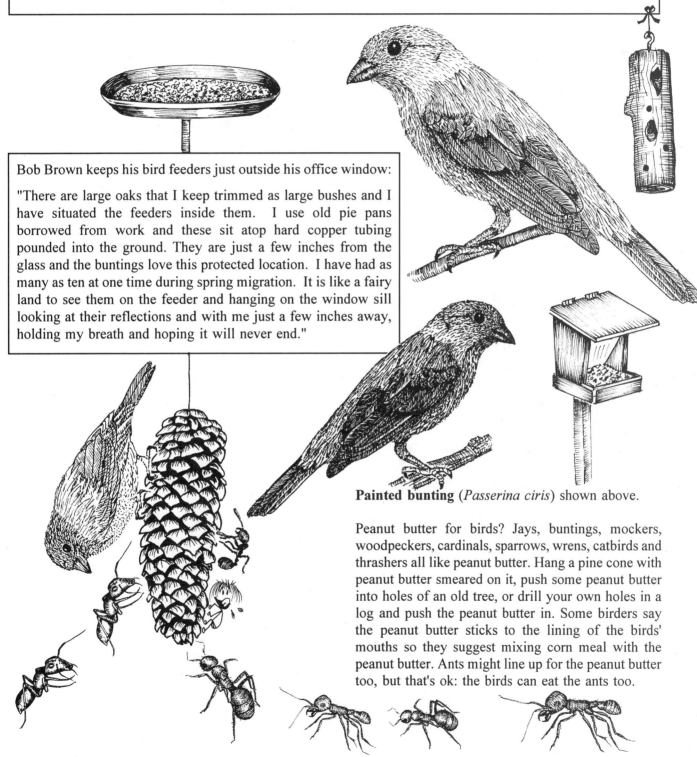

Bob Brown keeps his bird feeders just outside his office window:

"There are large oaks that I keep trimmed as large bushes and I have situated the feeders inside them. I use old pie pans borrowed from work and these sit atop hard copper tubing pounded into the ground. They are just a few inches from the glass and the buntings love this protected location. I have had as many as ten at one time during spring migration. It is like a fairy land to see them on the feeder and hanging on the window sill looking at their reflections and with me just a few inches away, holding my breath and hoping it will never end."

Painted bunting (*Passerina ciris*) shown above.

Peanut butter for birds? Jays, buntings, mockers, woodpeckers, cardinals, sparrows, wrens, catbirds and thrashers all like peanut butter. Hang a pine cone with peanut butter smeared on it, push some peanut butter into holes of an old tree, or drill your own holes in a log and push the peanut butter in. Some birders say the peanut butter sticks to the lining of the birds' mouths so they suggest mixing corn meal with the peanut butter. Ants might line up for the peanut butter too, but that's ok: the birds can eat the ants too.

"Ants in the house seem to be, not intruders, but the owners. Of all things they seem the least aware of human beings. Even a cockroach will scuttle at sight of the mistress of the kitchen, but a colony of ants goes ahead with its thievery under her eye and fury."
- Marjorie Kinnan Rawlings in *Cross Creek*

Hummingbirds

From "Nature Notes" in *The Limpkin* by Shirley Hills:
"One of the brightest events of the winter is watching the male ruby-throated hummingbird that visits the large clump of shrimp plants growing on the edge of the patio. During the early part of the winter he spent most of his time around the red flowers in the front yard but after the frosts wiped out most of the blooms I wondered if he would stay. The shrimp plants were not yet in blossom but this knowing little bird would come by everyday to check the condition of the buds ... The shrimp plants came through all the frosts unscathed and started blooming profusely. It was only then that I was able to get a good look at this little blur of energy as it often stood on one flower in a tightly packed cluster of blooms in order to reach a choice flower ... The warm spring days have turned the male ruby-throated hummingbird into a feisty busybody. He scolds me when I walk out of my laundry room door (which happens to be next to a clump of his favorite shrimp plant)."

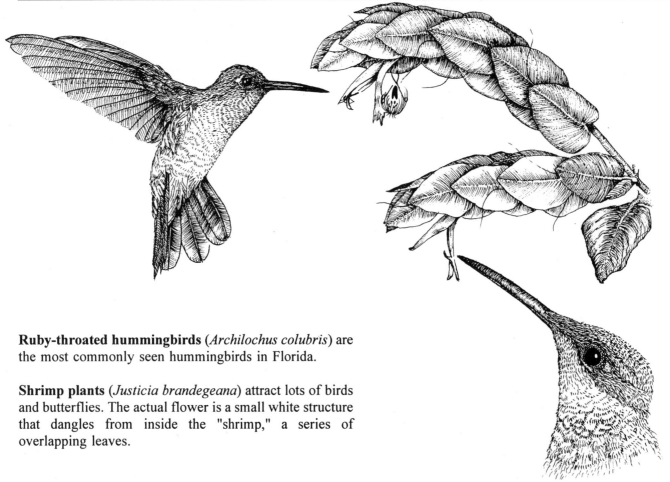

Ruby-throated hummingbirds (*Archilochus colubris*) are the most commonly seen hummingbirds in Florida.

Shrimp plants (*Justicia brandegeana*) attract lots of birds and butterflies. The actual flower is a small white structure that dangles from inside the "shrimp," a series of overlapping leaves.

In "A Passion for Hummers" from the February 1997 *Birder's World*, Judith Toups wrote, "Although the Mississippi River was long believed to be a barrier that most western hummers refused to cross, there were watchers waiting on the other side for the first wonderful discoveries of these jeweled vagrants ... Hummingbirds of western origin were already pioneering into the southeastern United States, mostly in fall and winter." At least 10 hummingbird species now come to Florida.

Toups adds, "Not only are the number and variety of hummers that winter in the southeastern United States causing birders to re-think what we thought we knew, but we are learning that what might have once seemed impossible is anything but." So, don't take bird guides as gospel. The birds I see through my binoculars don't always read the bird guides.

"Beware of any activity that requires the purchase of new clothes."
- Henry David Thoreau

Common Backyard birds

In the fall, I hear large flocks of noisy **blue jays** (*Cyanocita cristata*) gathering in the neighborhood near water and food feeders. But their autumn noisiness is classical music compared with the heavy metal of the spring: loud, raucous, and demanding. And as quickly as the music starts, it stops. Then a long eerie silence follows. During this quiet time, mated pairs skulk noiselessly through low bushes looking for building material to begin assembling their nests. Listen closely to hear the only sounds they make during this approach to incubation: simple soft calls between pairs, mostly the female accepting food from the male.

Cardinals (*Cardinalis cardinalis*), more tolerant than the jays, politely wait their turn for the feeder. I know their turn has come when I hear the *crack, crack* as they open sunflower seeds. Sometimes I see a male carry seeds to his nest mate. In Florida, nesting can go on from February through November. This creates a lot of activity for the male who carries food to his nest while still teaching an earlier batch of chicks their living skills.

Male and female **mourning doves** (*Zenaidura macrora*) look alike, with a distinctive black dot on the cheek and pinkish-orange feet.

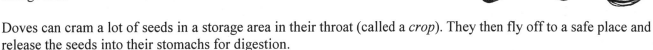

Doves can cram a lot of seeds in a storage area in their throat (called a *crop*). They then fly off to a safe place and release the seeds into their stomachs for digestion.

Doves have a slow take-off so they're easy prey for hawks and cats. But one adaptation that protects them is that their feathers come out easily; when the dove is attacked, the predator is often left with only a mouth full of feathers.

The soft cooing sound is from the male attracting a mate. His courtship lasts through the spring and summer. He struts in front of the female with his feathers spread, cocking his head up and down with a male bird's arrogance. Doves have more broods than our other neighborhood birds -- up to six per year. The female throws a sloppy nest together within a few days with careless abandon. Sometimes, if she's motivated, she lines the nest, but most of the time she simply lays her eggs on a loose arrangement of twigs.

Why do mourning doves squeak like rusty hinges when they take off? The high-pitched sound, made by their feathers as they flap, is a way to distract or frighten predators that may be lurking.

"We need above all, I think, a certain remoteness from urban confusion."
- Marjorie Kinnan Rawlings in *Cross Creek*

Escapees?

Sunflowers (*Helianthus annuus*) attract lots of birds, particularly **wild parrots**. In my neighborhood, flocks of noisy, bright green parrots called **blue-capped conures** (*Aratinga acuticaudata*) fly overhead, screaming and swooping like crazy dive bombers. Sometimes they stop to eat at yards with sunflowers. Parrot experts describe these conures as spunky and comical. Most of the wild parrots we see in Florida are imported birds that have escaped. Bird authorities inform me that they are known as "escapes," not escapees, as I used to call them.

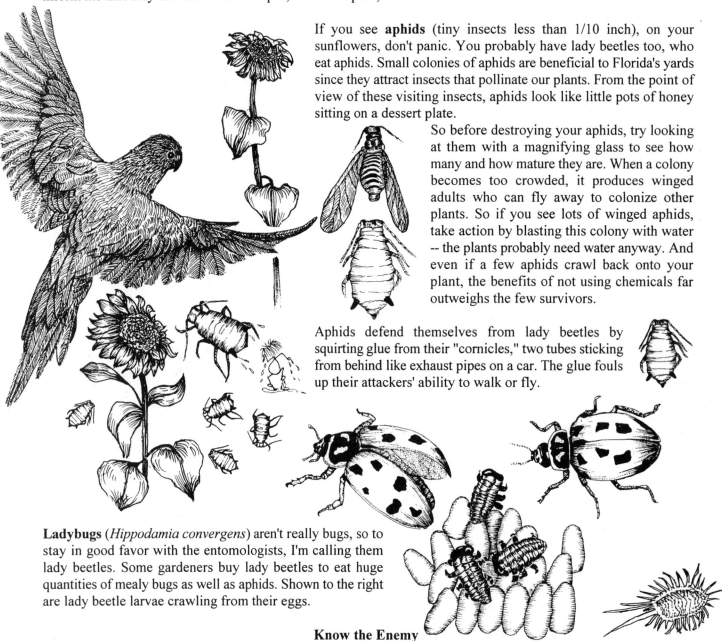

If you see **aphids** (tiny insects less than 1/10 inch), on your sunflowers, don't panic. You probably have lady beetles too, who eat aphids. Small colonies of aphids are beneficial to Florida's yards since they attract insects that pollinate our plants. From the point of view of these visiting insects, aphids look like little pots of honey sitting on a dessert plate.

So before destroying your aphids, try looking at them with a magnifying glass to see how many and how mature they are. When a colony becomes too crowded, it produces winged adults who can fly away to colonize other plants. So if you see lots of winged aphids, take action by blasting this colony with water -- the plants probably need water anyway. And even if a few aphids crawl back onto your plant, the benefits of not using chemicals far outweighs the few survivors.

Aphids defend themselves from lady beetles by squirting glue from their "cornicles," two tubes sticking from behind like exhaust pipes on a car. The glue fouls up their attackers' ability to walk or fly.

Ladybugs (*Hippodamia convergens*) aren't really bugs, so to stay in good favor with the entomologists, I'm calling them lady beetles. Some gardeners buy lady beetles to eat huge quantities of mealy bugs as well as aphids. Shown to the right are lady beetle larvae crawling from their eggs.

Know the Enemy

From Tom MacCubbin in *Florida Home Grown 2: The Edible Landscape*
"Handpicking insects from plants is an age-old pest control that still works fine today. It seems silly to spray a whole plant trying to kill a small caterpillar. Even a cluster of the pests on one plant out of five hardly warrants the effort of priming the sprayer. Handpicking is also a good way to get to know the bugs ... Some insects and related pests, especially slugs and snails, can be captured where they hide. Give them a false sense of security by providing places for them. Leave a board or a pot in the garden path - they like the moist, cozy darkness found underneath, Then take an early morning stroll to collect and catch and do them in."

"What lies behind us and what lies before us are tiny matters, compared to what lies within us.   -Ralph Waldo Emerson

Urban Pollinators

Letting native plants grow is a good way to attract pollinators such as moths, beetles, bees and bats. **Spanish needles** (*Bidens alba*) provide nectar and pollen for the pollinators. Spanish needles are also good hosts for insects not normally thought of as pollinators -- flies, mosquitoes and wasps.

How important are insects anyway? Without bees, wasps, flies, butterflies and moths, we'd have no wildflowers, citrus, or berry bushes. Messy gardens and yards are sanctuaries for pollinators so they can rest and nest under twigs and piles of leaves.

If you want the benefits of butterflies in your backyard, you'll need to accept their babies, the caterpillars who eat your plants for nourishment. Different butterflies are attracted to particular plants, so if you have a preference for a certain color or shape butterfly, you can choose the plant to attract those particular butterflies.

To choose plants for a butterfly yard, consider three qualities important to butterflies: color, scent, and surface convenience.

Lantanas (*Lantana involucrata*) make it easy for butterflies to feed because the clusters of brightly colored flowers give plenty of surface area. Consequently, the butterfly doesn't use a lot of energy to get from one flower to another. (Lantanas shown to the right.)

Providing white flowers is important for male butterflies who need a chemical found in white flowers to produce sperm.

Also important to consider is that certain butterflies only feed on specific plants; monarchs only feed on milkweeds; zebra longwings only on passion vines; zebra swallowtails only on pawpaws.

White flowers and lantanas both attract the **gulf fritillary** (*Argraulis vanillae*), one of the most common urban butterflies in Florida all year long. They feed on a variety of passion vines. Their beautiful wings are bright orange with silver white dots on black spots. (Gulf fritillary shown above.)

Fritillaries scatter their small yellow eggs, one by one, by flitting from passion vines to fences or posts to lay an egg here and there. When they migrate south, they fly about 10 miles per hour. When they come to a building, they fly up and over instead of going around.

"The soul should always stand ajar, ready to welcome the ecstatic experience."
- Emily Dickinson

13

Zebra Longwings

The zebra longwing was voted Florida's official butterfly in 1996. One of the most intriguing plant-butterfly relationships is between the **zebra longwing** (*Heliconius charitonius tuckeri*) and its host plant, the **passion vine** (*Passiflora* spp).

Passion vines produce poison to keep insects away, but zebra longwings have developed an immunity to the plant's poison. Not only is the zebra immune, it actually uses the poison to its own advantage to keep predators away: longwings taste horrible and any bird foolish enough to eat one will spit it out right away. (Most birds have learned to avoid zebras, but once in awhile I see evidence of their close encounters by seeing the butterflies with frayed wings.)

But Who's Fooling Whom?

But now the plot thickens: the passion vine needs some defense against the zebra longwing, so the vine has developed a way to attract ants which are the butterflies' greatest enemy. Two glands on the plant secrete a sweet liquid which attracts ants.

Adult zebra longwings don't eat the passion vine leaves, but rather the nectar and pollen (which is unusual since no other butterfly is known to eat pollen).

Female zebra longwings lay eggs *only* on passion vines. The eggs are keg-shaped and bright yellow with vertical grooves. The egg casings are eaten by the larvae after they hatch. The larvae continue to grow by feeding on the passion plant.

The chrysalis, an early development stage of this butterfly, has also developed a defense of its own: surprisingly, it makes a rasping sound when disturbed.

Zebra longwing caterpillar shown to the left.

Zebra longwings gather communally at night, always returning to the same perch to sleep. They sleep so soundly that you can pick one up and return it later without waking it or any of its roost-mates.

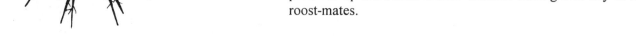
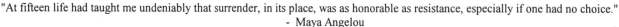

"At fifteen life had taught me undeniably that surrender, in its place, was as honorable as resistance, especially if one had no choice."
- Maya Angelou

Attracting More Butterflies and Moths

Did you ever notice how *fickle* butterflies are when they fly from flower to flower, changing their minds as if they can't quite find the exact right one? Evidently, if they're not looking for nectar plants for food, they're looking for an appropriate place to lay eggs. When you see one hovering over a leaf surface, you know she's searching for an egg-laying spot ... her fickleness is her way of smelling and tasting for the right chemicals. The larvae need specific plants for nourishment. Butterflies have organs on their feet that can taste food. They use vision to find nectar-producing flowers, but once they land on a plant, they rely on their feet to find the pool of nectar.

Drinking buddies

Sometimes you see a group of butterflies around puddles or wet areas: this communal gathering is called a *drinking club* or *puddling club* and is usually all males. While drinking, they become so absorbed that you can approach them without their noticing you. You can create a nice puddling area by filling a birdbath or dish with gravel or pebbles. This provides different levels for the butterflies to puddle.

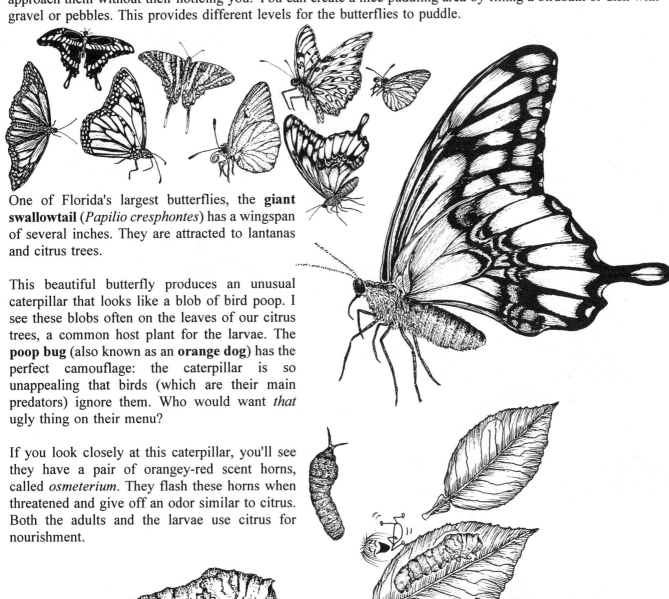

One of Florida's largest butterflies, the **giant swallowtail** (*Papilio cresphontes*) has a wingspan of several inches. They are attracted to lantanas and citrus trees.

This beautiful butterfly produces an unusual caterpillar that looks like a blob of bird poop. I see these blobs often on the leaves of our citrus trees, a common host plant for the larvae. The **poop bug** (also known as an **orange dog**) has the perfect camouflage: the caterpillar is so unappealing that birds (which are their main predators) ignore them. Who would want *that* ugly thing on their menu?

If you look closely at this caterpillar, you'll see they have a pair of orangey-red scent horns, called *osmeterium*. They flash these horns when threatened and give off an odor similar to citrus. Both the adults and the larvae use citrus for nourishment.

"No one can make you feel inferior without your consent."
Eleanor Roosevelt

15

Travelers

Monarchs (*Danaus plexippus*) live just long enough to travel thousands of miles each fall to mate in the tropics from as far away as Canada. Around October they pass through Florida on their way to Mexico. During their 3-week journey, they can navigate to the same trees their ancestors used (without ever having been there before).

Four-inch wingspans give monarchs the ability to fly up to 25 miles per hour. Their range determined by availability of their food, **milkweeds** (*Asclepias* spp.), which flourish everywhere.

Monarch flocks are always seen migrating *south*, but never *north*. How do they get back to Canada? Monarchs band together for their southward journey and stay together through the winter. But in the spring, they go back to being loners. On their return flight, they meander alone, free and easy with no social obligations.

During this gradual northward return, females lay eggs: green dewdrops, one or two at a time until each has laid 200 to 300 eggs, all underneath milkweed leaves. The larvae typically feed *only* on milkweed which makes them poisonous to predators. Their bright color is an advertisement of their noxious taste. Recently, however, some monarchs have been seen feeding on dogbane (*Apocynum* spp.) and these monarchs are harmless. Birds still avoid them, however, conditioned by bad memories.

Male monarchs have a conspicuous black dot (which is a cluster of scent scales) on the hindwing. This dot emits a fragrance to attract females.

On the Wing of a Question by Patricia Ryan Frazier

O' caterpillar, how did you guess
*   that it was time to now digress*
*   and change your pudgy form -- egress*
*   your mode of motion and of dress.*

To cease to sup on leaves, and why
*   did you turn your gaze upon the sky,*
*   then wrap in fine-spun coffin, die ...*
*   to arise again as butterfly?*

Moths or Butterflies?

What's the difference between moths and butterflies? The most obvious differences are that moths fly at night and have fatter, hairier bodies. Another difference is their antennae: butterflies' antennae end in a wider club shape, but most moths have feathery or slender antennae tips. (See the luna moth antennae on page 53.)

Long-tail skippers (*Urbanus proteus*) are medium-sized and brown with very long tails and green iridescence inside their wings. Are skippers considered moths or butterflies? That's unclear -- some professionals classify them with moths because of their moth-like features: small, thick, hairy bodies and hooked antennae. But typical of butterflies, they dance in the air when the sun is shining.

Oleander moths (*Syntomeida epilais jucundissima*) are exotic-looking, small, polka-dotted moths that come around each spring. Their larva is an orange caterpillar with long black hair tufts lined in rows over their body.

I see this caterpillar crawling, in its self-absorbed way, on leaves of **oleanders** (*Nerium oleander*). All parts of an oleander are poisonous, including the stems, sap and leaves, but the oleander caterpillar eats them with no ill effects.

Shown to the right is **plumbago** (*Plumbago auriculata*), the only blue flower I find in our neighborhood. Because of the unusual color, it attracts butterflies and insects that we normally don't see.

"This is the true joy in life, the being used for a purpose recognized by yourself as a mighty one, the being thoroughly worn out before you are thrown on the scrap heap; the being a force of Nature instead of a feverish selfish little clod of ailments and grievances, complaining that the world will not devote itself to making you happy."
-George Bernard Shaw

Plants to Attract Backyard Wildlife

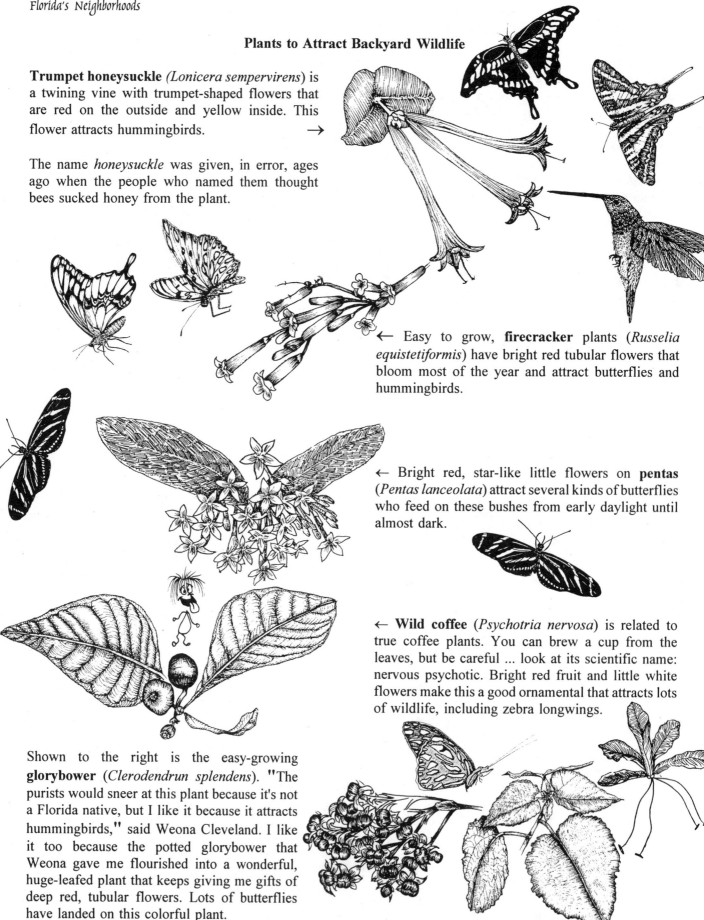

Trumpet honeysuckle *(Lonicera sempervirens)* is a twining vine with trumpet-shaped flowers that are red on the outside and yellow inside. This flower attracts hummingbirds. →

The name *honeysuckle* was given, in error, ages ago when the people who named them thought bees sucked honey from the plant.

← Easy to grow, **firecracker** plants (*Russelia equistetiformis*) have bright red tubular flowers that bloom most of the year and attract butterflies and hummingbirds.

← Bright red, star-like little flowers on **pentas** (*Pentas lanceolata*) attract several kinds of butterflies who feed on these bushes from early daylight until almost dark.

← **Wild coffee** (*Psychotria nervosa*) is related to true coffee plants. You can brew a cup from the leaves, but be careful ... look at its scientific name: nervous psychotic. Bright red fruit and little white flowers make this a good ornamental that attracts lots of wildlife, including zebra longwings.

Shown to the right is the easy-growing **glorybower** (*Clerodendrun splendens*). "The purists would sneer at this plant because it's not a Florida native, but I like it because it attracts hummingbirds," said Weona Cleveland. I like it too because the potted glorybower that Weona gave me flourished into a wonderful, huge-leafed plant that keeps giving me gifts of deep red, tubular flowers. Lots of butterflies have landed on this colorful plant.

A visitor once asked George Bernard Shaw why he kept no cut flowers in his home since he was so fond of flowers. "So I am," he said, "I'm very fond of children too, but I don't cut off their heads and stick them in pots all over the house."

Ants

Do ants have wings? Male ants and queens have wings: when the queen is ready to mate, she flies close to a male, and after mating in mid-air, the queen sheds her wings.

All worker ants are sisters and children of the queen. Males don't work -- their only purpose is to fertilize eggs to produce more females. (Male ants are produced from nonfertilized eggs.)

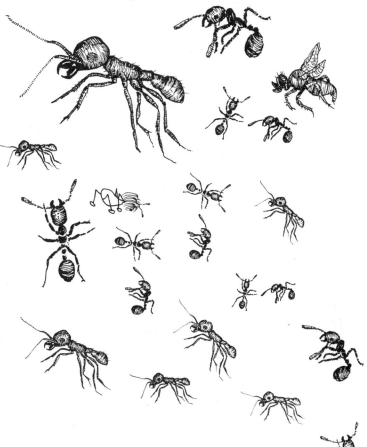

Fire ants in Florida are either native (*Solenopsis geminata*) or imported (*Solenopsis* spp.). Mean-spirited, fire ants clamp onto our flesh with their mouths and then inject venom with their rear-end stinger. We usually get bitten by more than one because the first biter emits an "alarm substance" to signal the others to follow. These ants are sightless, but they can find their way around through their sense of smell.

Pain in the Ants

Fire ants have been killing and causing pain since they came to this country on a ship from South America in the 1930s. For the first time since then, however, their reign of pain may be coming to an end. A South American phorid fly, a small, hunchbacked, two-winged insect, can destroy them by having their maggots eat the ants' heads. The adult phorid fly injects its eggs into the ant's body. When the maggots hatch, they crawl to the fire ant's head and release a destructive substance. About 12 hours later, the ant's head falls off. After fellow ants carry the head out to the colony's junk pile, the pupa matures into an adult fly and eventually emerges through the mouth of the dead ant. Sanford Porter, a researcher with the Gainesville branch of the USDA, claims the phorid flies won't attack anything but South American red fire ants.

Don't Kill the Messenger

← **Pharaoh's ants** or little sugar ants (*Monomorium pharaonis*) are tiny pale brown or reddish insects that nest in woodwork. Follow their trail to see where they live. Killing the thin parade on your counter won't get rid of the problem -- they're only the messengers, and the rest of the huge colony is safe, far back in a hard-to-get-to-spot.

Kill the Queen

Florida carpenter ants (*Camponotus floridanus*) (shown below) have reddish bodies with a black abdomen. They are often more than ½ inch. In houses, these ants tunnel in wood. Outside, they live in dead trees. Their well-kept nests hold thousands of ants, so spraying one or two in your house is like believing you'll destroy a beehive by swatting on the bee on your arm. Instead, follow an ant to its home. Destroy the entire nest without using chemicals by shooting water into their home. Flooding the nest will kill the queen, thus destroying the colony.

"There's an awful lot of irrationality in the attitude toward, and hatred of, fire ants. It's just another creature trying to make a living."
- Dr. Walter Tschinkel of Florida State University

Moles and Miners

From Tom MacCubbin in *Florida Home Grown*:
"Many animals simply meander into the lawn and garden looking for food. In their search for meals, moles raise the soil above the ground as they tunnel, and armadillos dig holes that are often a foot or more deep. Both are searching for grubs, earthworms and similar insects, their favorite foods. Gardeners can just ignore moles. They feed on harmful insects and aerate the soil. Moles move on when the food supply runs out."

Eastern moles (*Scalopus aquaticus australis*) eat insects and worms rather than grass or plants, so they don't destroy garden crops as some people think. They eat constantly because of their high metabolic rate, so you know they're eating a lot of insects we probably don't want. Let them be.

Moles have huge, powerful front feet made for digging long tunnels. They can burrow underground within 5 seconds. Part of their tunnel is another interconnecting series of deeper tunnels where they spend their solitary life. They only socialize once a year to mate and raise a litter. The network of tunnels also serves as a drop trap for insects like beetles who fall into the moles' tunnel while digging their own.

Moles make me think of me of **mole crickets**. (Also see page 58.)

Another strange creature who makes tunnels is the **leaf miner** (*Liriomyza* sp.). Leaf miners are fly maggots. They eat their way *in between* the membranes of a leaf, forming that familiar squiggly design we see on citrus leaves and citrus fruit. Adult leaf miners are about 1/20 inch long.

From Tom MacCubbin in *Florida Home Grown 2: The Edible Landscape*
"Newcomers often assume citrus trees are native. How else could the fruit do so well? In Florida, it is the most carefree fruit imaginable. A few blemishes will have to be tolerated, but the trees really don't have to be sprayed. Before the freeze of 1983, almost every street corner in Central Florida had its productive citrus trees. How could the Spanish, who introduced the trees, ever have known how well the crop would flourish? ... Citrus trees need no mollycoddling to produce sweet, juicy, home-grown fruit."

"Home life as we understand it is no more natural to us than a cage is natural to a cockatoo."   - George Bernard Shaw

Flies: The Good, The Bad and Maggot Bigotry

I never thought I was prejudiced until someone pointed out my prejudice against bigots. I also judged certain creatures as undesirable, such as maggots (which are the larval stage of flies). Are maggots less worthy than butterflies because they don't have colorful wings to camouflage their wormy bodies?

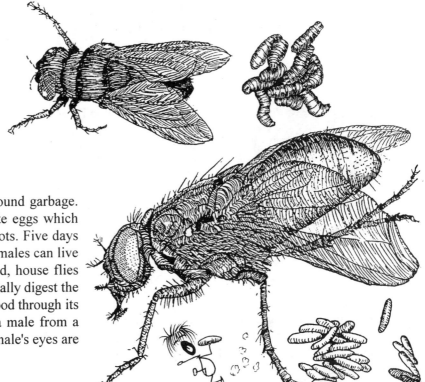

So what good are **green bottle flies** (*Phaenicia sericate*)? These ½-inch flies help disintegrate road kill. The flies lay their eggs on decaying matter, and for about two weeks of their larval stage they eat the dead flesh.

House flies (*Musca domestica*) hang around garbage. Females lay 5 to 6 batches of oval white eggs which hatch in 10 to 24 hours to become maggots. Five days later the maggots become adults. Adult females can live for about 26 days. To feed on solid food, house flies vomit on it first; the vomit's enzymes partially digest the food so the fly can soak up the liquified food through its sponge-like mouth. How can you tell a male from a female fly? The only difference is the female's eyes are farther apart. Like I'd notice.

Did you ever notice a house fly just sitting motionless on a windowsill and not moving even when you get close? Look closer and you'll see some white powder underneath the fly's body; this is from a spore that enters and grows inside the fly. When the spore is ready to release new spores, it has a way of attaching the fly to one spot. It then releases the spores from the fly's body until all that's left is an empty shell -- a mere shadow of its former self.

Ray Gant wanted to know if a *fly* doesn't have wings, is it called a *walk*?

Yellow flies (*Syrphidae* spp.), also known as hover flies (Also see page 35.) are smaller than house flies and have yellow bands around their abdomens. They have a bite more painful than a wasp sting, and it can take a week or more to heal. The digestive enzymes that these flies inject in you make your blood flow more freely. These enzymes are the cause of the sore.

God in His wisdom made the fly
And then forgot to tell us why.
- Ogden Nash

Annoying Neighbors

Tick Talk

Ticks (*Dermacentor variabilis*) (shown right) aren't insects, but are closely related to spiders and scorpions. After gorging on animal blood, a tick *will grow from $\frac{1}{8}$ inch to $\frac{1}{2}$ inch*. If you find a tick sucking on you or your pet, pull it out with tweezers, making sure to pull out the complete tick. If its mouth is left in, disease or infection may be transmitted.

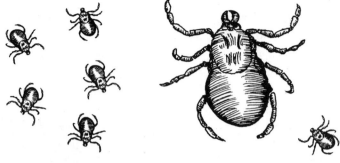

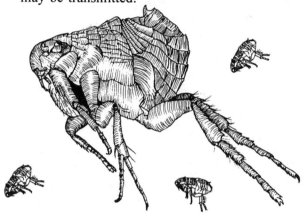

← **Fleas** (*Ctenocephalides* spp.) can live without food (animal blood) for up to 8 months. Dog fleas and cat fleas are not the same, but they both lay tiny white eggs about the size of this period ◡. The eggs develop into larvae that can survive on almost any organic material. That's how they can stay alive in our carpets for so long even when no animal is around.

Even though fleas evolved from flying insects, they have no wings and can't fly. So why did they lose their wings which took millions of years to evolve? Wings are vulnerable to destruction and because of a flea's lifestyle (cramped in between hairs), wings were simply a nuisance.

What are those annoying dust-like pests that swarm around my head like dirty clouds? When I swat them away, they scatter for a second and then come back to annoy me some more. Each tiny pest seems to zigzag in a frenzy competing with the other zigzaggers. **Gnats or midges** (*Tendipedidae* spp.) (shown right) are tiny mosquito-like flies that dance in the air. Gnats use our bodies as landmarks where they gather to meet mates. Females are attracted to this mass dancing and enter the swarm where the males then compete to mate. Hence the frenzy.

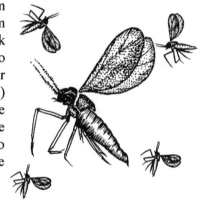

Love Bugs (*Plecia nearctica*) are harmless to humans since they don't bite or sting. But they do damage car finishes if you don't wash the insects off right away.

Love bugs indulge in massive mating flights in May and September for about four weeks -- that's when they crash and smash into our cars and leave the familiar mess. They mate anywhere -- on the ground, in mid-air, or on cars and buildings. Mating between a pair continues until the male dies. Females live about a week.

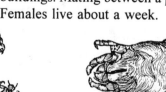

← **Dust mites** and pollen make me sneeze.

Shown above: **head lice**

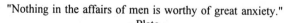

Little Things in and Around the House

Bedbugs (*Cimex lectularius*) (shown right) are 1/8-inch long. These tiny reddish-brown insects drink human blood, usually while the human is sleeping. After a bedbug has taken its first bite on you, it moves a little and takes another bite, then another. If you see a trail of tiny blood clots somewhere on your body when you wake up, they're probably from bedbugs. If you sneeze when you crawl into bed, it could be another indication that bedbugs live in your mattress. Nymphs take 6 weeks to a year to mature, and the adults can live without food for 15 months.

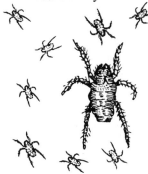

Shown to the left, **chiggers**, (*Eutrombicula alfreddugesi*) (also known as redbugs) are the tiny (*1/100 inch*) larvae of harvest mites. In the spring, when chigger larvae hatch, they want tender flesh for their food. If you happen to be walking in their habitat, the parasitic larva will attach its mouth to your skin, usually under the tightest part of your clothing. It then injects a digestive juice that disintegrates the skin cells which it eats. The skin area becomes red and swollen, engulfing the chigger and giving the wrong impression that the chigger has burrowed into your skin. Even after the chigger dies or falls off, its bite continues to itch for a few days.

Look closely

One advantage of never cleaning my house is I get to see wonderful creatures like **dangling spiders** (*Cyclosa bifurca*). (I couldn't find a common name for these fascinating spiders so I dubbed them *dangling* since they hang at the end of a series of dangling tiny egg cases.) I find them hanging from the porch ceiling and under the awnings. I used to think they were strange little dust clumps until I got my magnifying glass out one day -- and wow! -- was I amazed to see a beautiful pale spider at the end, camouflaged to look *exactly* like her egg cases.

Camouflage

Most spiders deposit their eggs all in one laying, but this dangling spider makes an elaborate structure of a string of egg sacs hanging from a disorganized web of silk threads. At the end of this strand, she imitates the color, shape, position, and size of the egg case by folding her legs flat against her body. If you disturb her, she'll run up the strand and seem to disappear again. Shown to the immediate right is an enlarged view of this amazing dangling spider. The actual size is about ⅓ inch.

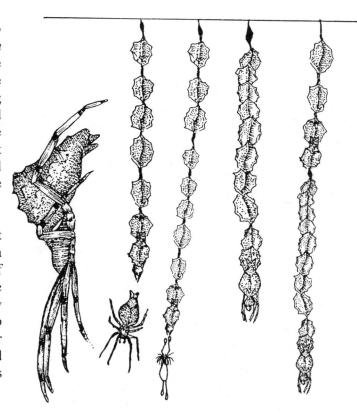

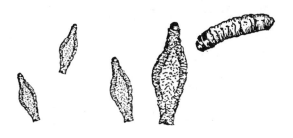

Plaster worms (*Phereoca walsinghami*) (shown left) are about ½-inch long and look like cantaloupe seeds clinging to my walls. I used to think these flat little things were flecks of plaster ... until I noticed one moving. I looked at it through my magnifying glass and YIKES -- I saw a tiny worm inside, poking its head out. I love when that happens.

♫"It's important to make someone happy. Make just one someone happy. Where's the real stuff in life to cling to? Love is the answer." ♫
- Jimmy Durante

ROADSIDES

Every fall, the hickory tree next to Dr. Beard's office in Melbourne Beach drops its fruit onto the parking lot, tempting our neighborhood squirrels. The nuts are too hard for the squirrels to crack, but these clever animals have figured out a way to get to this tasty food. Debbie Beyer, Dr. Beard's gentle dental assistant explained, "The squirrels wait until our patients drive off. They know the cars will crack the nuts open. Look, there's a squirrel now, waiting for a car to pull out." Sure enough, the watchful squirrel ran behind a departing car and grabbed the meat from a newly cracked hickory nut.

Squirrels

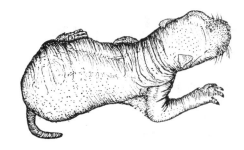

Squirrels (*Sylvilagus carolinensis*) often nest in palm trees, building large, loosely structured nests of leaves. Their babies are hairless blobs of gray flesh, and are small enough that several could fit in my hand.

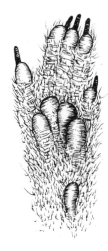

Squirrels' hind feet are long and sturdy. Five long toes on each foot make climbing, gripping and dangling from limbs easy. Their shorter front feet only have four toes, but are made to grasp nuts and seeds.

If squirrels didn't have those fluffy tails, we'd call them rats; if they had wings, we'd call them bats. Bats and rats have bad reputations, but squirrels are considered cute. Go figure.

Why do squirrels constantly flick those fluffy tails? A predator's attention will be drawn to the flicking tail and will attack the tail rather than the squirrel's vulnerable body. That's why we often see squirrels with missing or partially missing tails.

"The one fact that I would cry from every housetop is this: The Good Life is waiting for us -- here and now.
- B. F. Skinner

25

What is it? A palm tree or a strangler fig?

Many trees in my neighborhood have the trunks and leaves of strangler fig tree, but, strangely, palm fronds are growing from the middle. How can this be? When I discovered that the "outside" of this tree was a **strangler fig** (*Ficus aurea*), I realized that the tree had slowly, over the years, "changed" its identity from a cabbage palm to a strangler fig. A strangler fig is in the same family as banyan trees.

How does a strangler fig grow?

Fig seeds are carried by birds whose droppings splatter on the palms. The sticky seeds get lodged in the space between the frond stems, and begin to grow right there. When the seed germinates, it sends roots downward. Once the roots reach the soil, the fig's growth accelerates until it engulfs its host tree. The palm tree, which is still in the middle giving the fig support, eventually dies from starvation (rather than strangulation) since the strangler fig deprives the host tree of nutrients. Strangler figs can also grow from seeds in the ground, but these don't grow very tall without the support of a host tree.

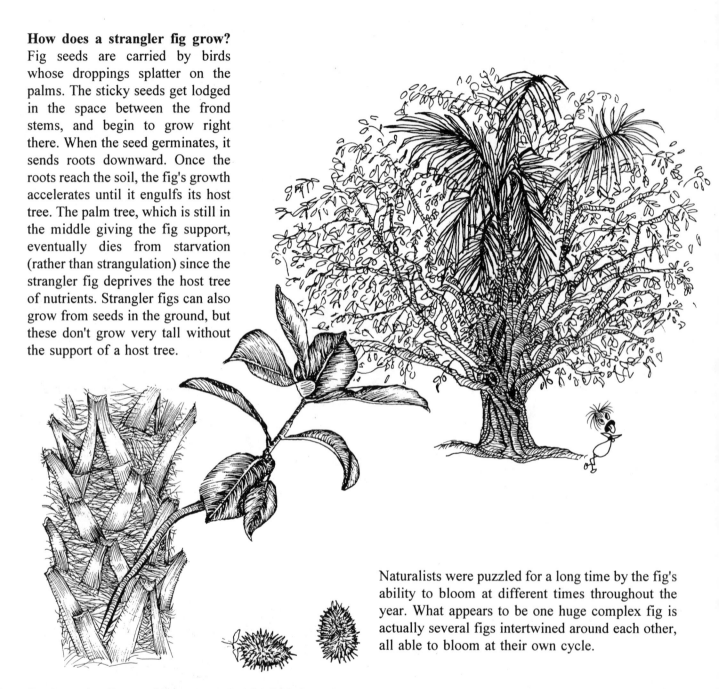

Naturalists were puzzled for a long time by the fig's ability to bloom at different times throughout the year. What appears to be one huge complex fig is actually several figs intertwined around each other, all able to bloom at their own cycle.

Seeds can be dispersed by water, wind, birds and mammals. When I walk around my neighborhood, I let my arms and shoulders brush against the trees and plants so that I can consider myself as a pollinator or seed carrier too. Sometimes it's painful, like when I get a sand spur in my foot. Seed pods such as burrs cling to animals because of tiny hooks on each pod. In 1948, Georges de Mestrel, a Swiss scientist found seed pods sticking to his clothing after a walk. He looked at the pod under a microscope and thought, "Velcro!" (He spent 8 years figuring that out.)

"Simplicity is the way we open to everyday wonder."
Jack Kornfield in *A Path with a Heart*

Non-native Trees: **Dead is Good**

Australian pines (*Casuarina equisetifolia*) were planted as a wind break in the 1920s. They were common until recent freezes. Now the (mostly) dead trees provide nesting habitat for birds such as woodpeckers. These trees are not encouraged (and not legal to plant) in Florida; very little can grow underneath them because of the thick bed of pine needles. In addition, their shallow root system doesn't give them stability during strong winds, causing serious damage when they crash to the ground.

Even though Florida is better off without Australian pines, the dead trees are useful to wildlife for nesting, hiding and feeding. Since Florida has lost a lot of its natural habitat, keeping the dead trees where they are helps the wildlife.

Woodpeckers love dead trees like Australian pines. Many insects live beneath the bark of these "lifeless trees."

Woodpeckers can get a good grip and keep their balance by using their strong toes. Notice the two toes in front and two sticking out backwards. Stiff tail feathers also add balance as the woodpeckers use them to push against the tree.

Pileated woodpeckers (*Dryocupus pileatus*) can be seen and heard in most neighborhoods because they have adapted to urban living much better than their relative, the (probably) extinct ivory-billed woodpecker. Pileated woodpeckers are large, colorful and noisy. Their bright red head is striped with white, making them one of the most easily identifiable of all our neighborhood birds. They also get my attention with their loud, raucous calls and monotonous drilling sounds.

Another non-native plant, the **Brazilian pepper** (*Schinus terebinthifolius*) was brought to Florida in the 1890s as an ornamental and was called Florida holly.

Brazilian peppers are salt-tolerant and fast-growing, but their aggressiveness blocks sunlight from lower plants and chokes out less aggressive native habitat. Their seeds are carried by birds, so dispersal is widespread and difficult to control. Because they are in the same family of plants as poison ivy and poison sumac, they cause skin irritations in some people. "Pepper Busters" in my neighborhood donate their time and energy to destroying these invasive plants.

"Extinct birds lay very few eggs." -A student's exam answer recorded in *More Anguished English* by Richard Lederer

27

Woodpeckers in NASA's Neighborhood

Another woodpecker, the **northern flicker** (*Colaptes auratus*), isn't welcome in many neighborhoods because they drum on gutters, windows and wooden doors, damaging wood as well as disturbing the peace.

At my work neighborhood, a problem with flickers became so bad that a special team was brought in to solve it. At Kennedy Space Center in 1995, the space shuttle Discovery was postponed because flickers had drilled hundreds of holes in the insulation of the external tank.

Why would woodpeckers bother pecking on a fuel tank? The light, brown insulation is similar in color to the birds' favorite palm trees. But the trees in nearby scrub habitat had been taken over by **starlings** (*Sturnus vulgaris*). The woodpeckers couldn't compete with the large flocks of starlings, so they searched for other sources for nesting cavities and food. The shuttle's fuel tank insulation was the right texture and color. But each time the birds drilled through 5 or 6 inches they would hit metal, so they had to keep moving a few inches away to try again. Eventually the insulation was covered in holes, endangering the shuttle's mission.

Solutions to keep the woodpeckers away from the shuttle, such as displaying owl balloons and using chirpers, streamers, and horns only worked temporarily. What could the woodpecker experts do? Move the starlings? Remove their food source? Add more scrub? Change the color of the insulation? Move the woodpeckers? Cancel the shuttle program?

The solution to the Shuttle's woodpecker problem was described by Linda Hoefer, operations security and a member of the Avian Deterrence Team at Kennedy Space Center: "We human beings are learning to be neighbors the animals. During non-nesting periods we leave them alone, but during nesting season we watch for the flickers to settle onto the fuel tanks. Then we use all the tools we have: owl balloons, chirpers, streamers, and horn blasters. Besides not harming the flickers, the horn blaster gives me an opportunity to let out my anger."

Starling shown above.

"Everything has its wonders, even darkness and silence, and I learn whatever state I may be in, therein to be content."
- Helen Keller

Roadside Birds

The **mockingbird** (*Mimus polyglottos*) is the official state bird of Florida. They are the most vocal birds in my neighborhood. Their scientific name means *imitator of many languages*. Their range of songs amazes me -- all those songs come from one bird? Are they imitating messages from other species? Or are they recording all the events of the neighborhood like our Town Historian, Frank Thomas, who records the history of our town?

I hear and see mockingbirds defending their territories with swooping dives at cats and screams at crows.

Mockingbirds maintain two different, clearly defined territories; one in the spring for breeding and one in the fall for food gathering. I hear the difference between the two seasons in their calls: in the spring, only males sing: a loud, creative, demanding, and versatile song. He gets quiet when he finds a mate. But in the fall, both sexes sing and their territory changes, centering around food rather than nesting. Their aggressiveness is aimed at food competitors: mostly other birds, including jays, starlings and other mockers.

Roadside birds, especially mockingbirds, are attracted to **beautyberries** (*callicarpa americana*).

Chipping sparrows (*Spizella passerina*) are common neighborhood visitors in the winter, poking around for seeds and insects. I hear them before I see them, recognizing their long single-pitched trill. They usually hang around under bushes and shrubs.

"In heaven an angel is nobody in particular."
- George Bernard Shaw

29

More Common Roadside Birds

American crows (*Corvus brachyrhunchos*) have a complex language of *caws*, *carr-carrs*, *cahs*, and *scat scats*. They have 23 different vocalizations including screams, cackles, rattles, sharp notes, fast notes -- all heard as warning signals, love messages, food alerts, or general reporting. Whenever I hear crow commotion in the backyard, I want to know what they're saying. Small talk? Birds don't waste their breath just to be heard -- each sound says something relevant.

An important function of imitating each other's calls is to strengthen the bonds between mated pairs and their group. They maintain strong social contacts. American crows can learn human words too (or is it just an ability to imitate?)

Fish crows (*Corvus ossifragus*) are true scavengers. They eat anything they can find or steal, which is why we see them at the beach, on streets, in vacant lots and in our backyards. The fish crow's call is a short nasal *cah cah*, different from the American crow's full-throated demanding *CAH*.

Fish crows are notorious for eating the eggs and young of other birds.

Common grackles (*Quiscalus quiscula*) have become more common in neighborhoods than in their own habitats of woods and farmland. They are easy to distinguish from crows by their purplish-brown heads and iridescent bronze bodies. They are also shorter and thinner than crows. Females are similar to males but duller. Grackles' eyes are pale, unlike crows who have black eyes. Sometimes they hang out with large flocks of crows.

Birds use ants to control parasites on their feathers and skin. Some birds crush ants on their bills and smear them on their skin like an ointment. Others simply sit on an ant hill and let the ants crawl through their feathers. Grackles use marigolds, which contain an insecticide (pyrethrum) to control parasites. Recently, British researchers witnessed a grackle preening itself with bits of lime that it had smashed. The researchers put some lime peel extract on some bird lice to see what would happen, and the lice died within seconds.

"In three words, I can sum up everything I've learned about life. It goes on." - Robert Frost

Roadside Predators

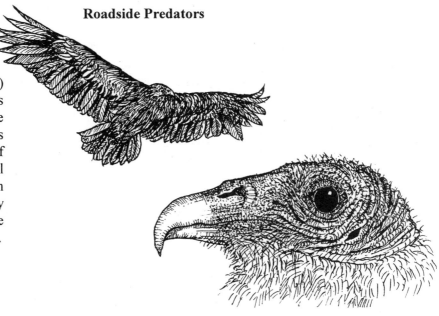

Turkey vultures (*Cathartes aura*) (shown right), are also known as turkey buzzards. They are the sanitation crew for Florida's neighborhoods. Their sense of smell is so keen, they can smell road kill from far away and within minutes after death. Lots of turkey vultures fly down from the northern states during the winter.

The **American kestrel** (*Falco sparverius*) (shown left) is about the same size as a blue jay. It's also called a sparrow hawk because it preys on LBJs (little brown jobs) like sparrows.

The kestrel is one of our most familiar urban birds of prey, commonly seen on telephone poles and wires during the winter.

Kestrels are mostly solitary even during breeding. For a few months during courtship, breeding and incubation, both males and females have marked divisions of chores; the female always stays near the nest, whether feeding or incubating, while the male stalks his hunting grounds for prey. He delivers his catch to the female and then takes off again. Interesting to note that their routine of hunting and receiving begins while the female can still hunt for herself long before incubation.

Red-shouldered hawks (*Buteo lineatus*) (shown right), are sometimes called chicken hawks. Although not as commonly seen as vultures and kestrels, they are often spotted in urban areas and along roadsides. They use fence posts, signs and telephone poles as look-out sites, perched for finding rodents, lizards, insects, frogs and small birds.

Red-shouldered hawks are a few inches smaller than their close cousins, red-tailed hawks, and have longer tails and wings.

"Your work is to discover your work and then with all your heart to give yourself to it."
- Buddha

Dandelions

The word **dandelion** (*Taraxacum officionale*) comes from the French *dent-de-lion* or *lion's tooth*. The flower is actually a cluster of many flowers. Pulling out dandelions as weeds and tossing them aside won't stop their life cycle. Seeds are carried away by the lightest puff of wind, and each seed has hook-like hairs (achene) at the end that help attach it to the soil.

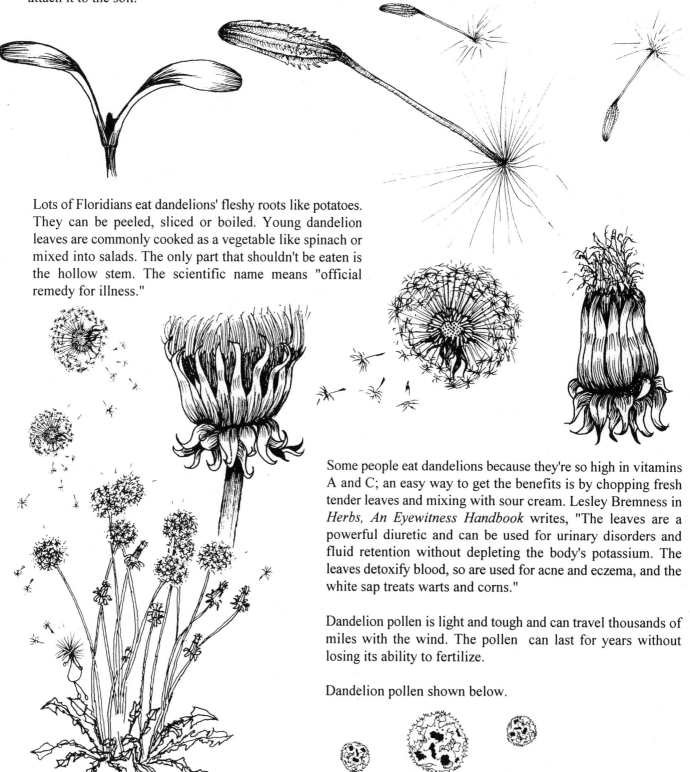

Lots of Floridians eat dandelions' fleshy roots like potatoes. They can be peeled, sliced or boiled. Young dandelion leaves are commonly cooked as a vegetable like spinach or mixed into salads. The only part that shouldn't be eaten is the hollow stem. The scientific name means "official remedy for illness."

Some people eat dandelions because they're so high in vitamins A and C; an easy way to get the benefits is by chopping fresh tender leaves and mixing with sour cream. Lesley Bremness in *Herbs, An Eyewitness Handbook* writes, "The leaves are a powerful diuretic and can be used for urinary disorders and fluid retention without depleting the body's potassium. The leaves detoxify blood, so are used for acne and eczema, and the white sap treats warts and corns."

Dandelion pollen is light and tough and can travel thousands of miles with the wind. The pollen can last for years without losing its ability to fertilize.

Dandelion pollen shown below.

"One of my favorite drinks is dandelion tea ... you certainly do not have to be sick before you drink it."
-In *Sybil Leek's Book of Herbs*

32

Weeds or Flowers?

⬎ Lantanas (*Lantana involucrata*)

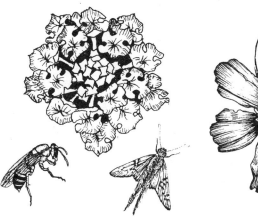

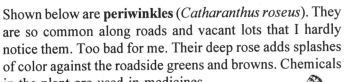

Shown to the left is the **coreopsis** or tickseed flower (*Coreopsis* spp.). The word *coreopsis* comes from the Greek *koris* and *opsis* which means "looks like bugs" because the seeds look like insects.

Shown below are **periwinkles** (*Catharanthus roseus*). They are so common along roads and vacant lots that I hardly notice them. Too bad for me. Their deep rose adds splashes of color against the roadside greens and browns. Chemicals in the plant are used in medicines.

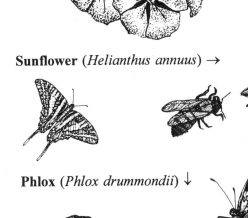

⬀ Black-eyed Susans (*Rudbeckia hirta*) are much more than just weeds to me. I can't believe they simply grow along roadsides and in vacant lots like they're no big deal.

Sunflower (*Helianthus annuus*) →

Phlox (*Phlox drummondii*) ↓

⬀ Spanish needles (*Bidens alba*)

"A weed is no more than a flower in disguise." -James Russell Lowell

Baggy and Stinky

I remember discovering my first **bagworm** (*Chridopteryx ephemeraeformis*) on the **ixora** (*Ixora coccinea*) bushes just outside our office building at Cape Canaveral Air Station. I was with our secretary Mary Witcher who, like me, thought these dangling bags were the cleverest decorations. We carefully examined the stick-structures, marveling at their intricate architecture. We didn't notice that the ixora leaves were more than half eaten because of these bag ladies.

Inside each of the stick houses is a worm-like creature. Look closely and you can see her head sticking out. Bagworms create camouflages by wrapping themselves in a silk bag and then covering this bag with twigs and bits of leaves from the host plant. Males eventually leave their bags to become moths, but the females never leave. The female has no wings, eyes, legs, antennae, or mouth -- only a soft, white, grub-like body whose sole function is to lay eggs in her bag.

Ixora or flame-of-the-woods (*Ixora coccinea*) has small orangey-scarlet flowers.

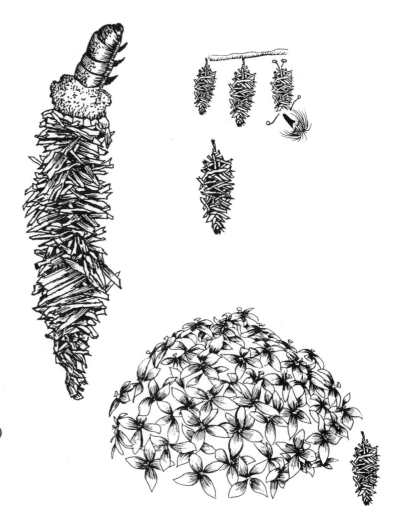

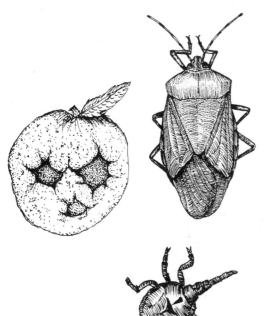

Southern green stinkbugs (*Nezara viridula*) are less than an inch long. They're also called shield bugs because of their body shape. Their pretty green bodies contradict their disgusting smell.

They harm fruit trees and vegetable plants by sucking the sap from pods, buds, leaves, seeds, and blossoms. When the stinkbug has sucked sap from a young fruit, "catfacing" will result: a deformity on the fruit's surface creates a cat's face when the fruit matures.

Other stinkbugs (brown and harlequin) are at least as damaging, but one stinkbug is a really nice one to have around. The **predaceous stinkbug** (*Euthyrhynchus floridanus*) attacks harmful bugs such as the oleander caterpillar. These stinkbugs are easy to identify by their dark blue bodies with orangey-red markings, long snouts, and smaller bodies. If you see this one, let it be.

People in New Guinea eat stinkbugs, relishing their crunchy almond flavor, and they're highly nutritious because of their protein.

"For the sense of smell, almost more than any other, has the power to recall memories and it's a pity that we use it so little."
- Rachel Carson

Killer Bees

The social structure of bees is sophisticated and complex. When you swat or slap a bee, it releases an alarm scent which signals hundreds of helper bees to come to the rescue. And during mating, a queen bee flies high in the air, with a bunch of drones chasing after her. The queen will only mate with the one who is strong and fast enough to reach her.

What's going on with those killer bees? Killer bees look like our honey bees except they're slightly smaller, and they have shorter tongues and forewings. One identifying characteristic is their nervous and erratic behavior. Because of the problem in identifying killer bees, it is hard to know where they're headed. Determining their direction is particularly difficult since they are mingling with our gentler honey bees.

How did killer bees get to the United States?

In the 50s, African queen bees (killer bees) were brought to Brazil for research because, even though they had mean spirits, they were wonderful honey producers. Accidentally, some bees escaped. By 1957 the bees had traveled north through South America toward the States, their fiery tempers and venomous stings causing cattle deaths along the way. The killer bees spread farther north and by 1990 they were discovered in Texas. Predictions that the killer bees would arrive in Florida by 1996 didn't prove true. But Florida is a perfect host state for killer bees because they thrive in this tropical climate.

Lawrence Pringle in his book *Here Comes the Killer Bees* writes several accounts of incidents that gave the killer bees their mean reputation: "One man was found dead with a thousand stings on his head; he had shot himself to end his agony. Another man, on horseback, collided with a swarm of Africanized bees. The horse threw him to the ground, breaking the man's leg, and then ran off with the bees in pursuit. Three days later the horse died from the countless stings it had received."

The sting of a killer bee is no worse than our honey bee's sting, but several hundred tiny doses of venom can be fatal. A swarm of African bees killed a rabbit in 30 seconds.

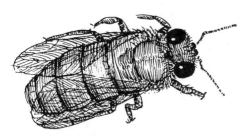

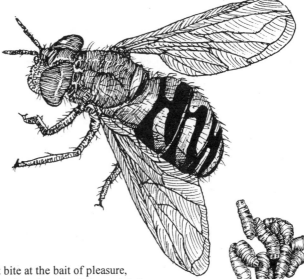

Hover flies or syrphid flies (*Syrphidae* spp.) (shown to the right) resemble bees and even visit flowers in the same way as bees do, feeding on nectar. These bee-imitators don't bite or sting and are good pollinators. Their offspring, green maggots, feed on aphids.

"Do not bite at the bait of pleasure,
til you know there is no hook beneath it."
- Thomas Jefferson

35

The Bees

More trouble for our honey bees? Ninety percent of the nation's wild honey bees have been killed by an exotic mite, *varroa*. The immediate impact is that plants that rely on honey bees for pollination won't be available as food for migrating birds. Also, small rodents that depend on seeds will suffer. And then hawks, owls and snakes will miss their food sources. One hope is that commercial bee colonies will be able to develop a bee strain that can survive the mites. Managed bee colonies can split off in the spring to find homes in natural cavities. Another hope is that carpenter bees and bumble bees may replace the missing honey bee as pollinators. Until then however, "the whole ecology of natural systems that rely on bees for pollination will have serious problems." - Tom Rinderer, Director of USDA's Honey Bee Breeding and Genetics Laboratory in Baton Rouge, Louisiana.

Florida has lots of different types of bees and the way the experts tell them apart is by their wing *venation* (the arrangement of veins in their wings).

What's the difference between a bumble bee and a honey bee?
Bumble bees are much larger than honey bees and have fuzzy bodies. Also, Rhonda Theobald says, honey bees make honey while bumble bees only bumble.

American bumble bees (*Bombus pennsylvanicus*) are social and well organized. They are more aggressive than honey bees, and they have a powerful sting. Unlike honey bees who only sting once, bumble bees can sting many times. They live in abandoned places like cars and old furniture. Like honey bees, they go for nectar, but they don't produce honey. But they're great pollinators because they have very long tongues that can reach into deep blossoms that honey bees can't reach.

Bumble bees pollinate by adjusting the frequency of their buzz -- changing their buzz so pollen shakes off the plant from the vibrations. Their hind legs have pollen baskets, specialized hairs used to collect pollen.

Honey bees (*Apis mellifera*) are good pollinators and also have pollen baskets on the upper part of their hind legs. A pollen comb, which is lower on the leg, brushes pollen from the body into the basket. Honey bees are more interested in nectar than in you, so they'll usually only sting when threatened. But, one bee can signal the rest in the hive to swarm. When they sting, they inject venom with their stingers which stays in your skin with the venom sacs. The stinger has a poison gland with sharp barbs to keep it firmly in your skin; as the stinger remains in your skin, venom continues to pump. (The bee dies after it has lost its stinger.)

From Bill Zak in *A Field Guide to Florida's Insects*
"The honeybee is everybody's friend. She is wrongfully blamed for many stings that are actually performed by the aggressive "yellow jacket" paper wasp. This popular insect is the prime practitioner in the insect world of the old philosophy ... Live and Let Live."

"Aerodynamically the bumble bee shouldn't be able to fly, but the bumble bee doesn't know it so it goes on flying."
- Mary Kay Ash

Wasps

Justin O. Schmidt is an entomologist with the U.S. Department of Agriculture. He has developed a way to measure pain associated with stinging insects. In his Pain Index Chart, he describes a **yellow jacket** (*Vespula* spp.) sting as, "Hot and smoky, almost irreverent. Imagine W. C. Fields extinguishing a cigar on your tongue." Yikes.

A yellow jacket (shown to the right) will sting at the least provocation. They hang around vacant lots in ground nests. Most yellow jackets die in the winter, except the females who spend the winter in protected soil or leaf litter.

About the sting of a **southern paper wasp** (*Polistes* spp.), Justin Schmidt says, "caustic and burning with a distinctly bitter after taste. Like spilling a beaker of hydrochloric acid on a paper cut."

Paper wasps (shown to the left) build their nests from bits of bark (or paper), creating an artistic array of colors in their intricate nests. The wasp chews the bark into pulp and mixes it with saliva until it becomes wet paper, like when I make papier-mâché. (However, I don't have enough spit, so I use water.)

I see their nests under soffits and ceilings and on house walls.

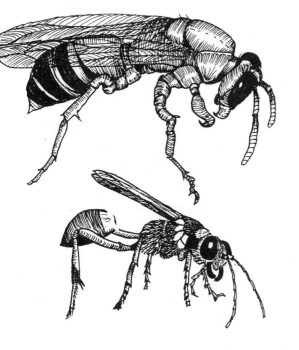

Some wasps have a complex social structure like yellow jackets, but others, like the **mud dauber** (*Sceliphron caementarium*) (shown left) are solitary. Their heavy jaws are used to shape and carry mud which they use to build their nests. They build their nests just about anywhere -- on walls, windows, plants, ceilings. When the nest is complete, the female fills each cell with a spider body which she paralyzes with her stinger. She then lays an egg in each spider-filled cell and seals it off with a dab of mud. Her job is complete, and she takes off. During their maturation, the larvae feed on the paralyzed spiders.

Adults emerge from their cells in late spring or early summer. They are about an inch long with dark metallic bodies and yellow marks.

"What a wonderful life I've had! I only wish I'd realized it sooner."
-Colette

SCRUB

"The scrub is the most ancient of Florida's natural communities. It holds many biological treasures distinctly Floridian, but so little known. Half of its plants and animals are unique, found nowhere else on the globe ... Scrub is subtle, gnarled, sometimes impenetrable. Scrub is glare and prickle. Its plants are clothed in the khaki of camouflage; its animals are dry-skinned and fossorial. Its shaman is the sky-blue scrub jay, and there is no green more hopeful than the scrub oak in springtime."

— Susan Cerulean, Florida wildlife author

What is scrub?

To see scrub from the road, you might think, "Oh yeah, that tangled mess of dried up weeds." But, whoa! Look again. Inside is a complex community of scrub jays, gopher tortoises, gopher frogs, pocket gophers, lizards, mice, snakes, beetles, lichens, mints, spiders and oak trees. Scrub is Florida's unique enchanted forest of secretive creatures and hidden agendas; a neighborhood of wildlife whose peculiar nature we're just beginning to uncover.

When asked "Why bother saving this dried up old scrub?" Florida scrub researcher Dr. John Fitzpatrick answered: "Why save the Mona Lisa? It's old and doesn't produce anything for us either ... Scrub is a gold mine of genetic information. These are plants that have spent a million years evolving ways to deal with a very specialized problem: hot, wet summers, cool, dry winters, poor soil - all things that humans have to deal with ... These plants may be holding secrets to how we can end up genetically engineering our agricultural systems to improve them, especially in dealing with a warmer planet."

Several birds are associated with scrub territory, some as predators like kestrels, loggerhead shrikes, red-tailed hawks and Cooper's hawks, and some as regular visitors like rufous-sided towhees, ground doves, mourning doves and mockingbirds.

But the most exemplary birds of the scrub are the **scrub jays** (*Aphelocoma coerulescens*).

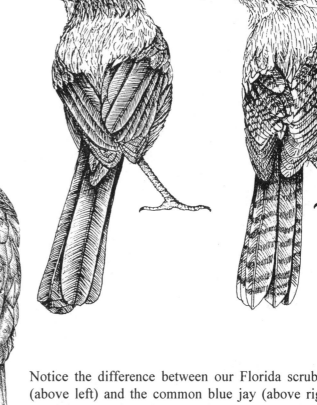

Notice the difference between our Florida scrub jay (above left) and the common blue jay (above right). Scrub jays are about the same size as our more familiar blue jays, but their color is more subtle with various shades of gray. A white eye-line blends with a frosted forehead. Stripes on the throat continue down along the breast, and a thick blue necklace matches the blue of the wings and tail. Blue jays have a crest on the back of their heads which the scrub jays lack.

"Here is a community of rare animals living on historic shorelines and dunes left from when the sea levels dropped millions of years ago. Only a fraction of this land is left now. Give the scrub jays half a chance and they will make it. Half a chance may be all they'll have."
- Margaret Broussard, Founder of Friends of the Scrub

Scrub Birds

Scrub jays are the friendliest birds I've ever known. My first experience with a scrub jay was *feeling* one perch on my head. One day in the 1980s, while I was I bicycling with my husband Gary, we passed a scrub lot in Melbourne Beach. I had never paid attention to this scrubby patch of gnarled and tangled weeds before, but now! Now a gentle bird from these "weeds" was sitting on my head! This friendly bird then jumped from my head to my handlebars, and looked at me, as if to say, "Well? How about some food. Please?" We raced home to get peanuts and birdseed. When we returned minutes later, the scrub jay was hopping on the hot, dry, sandy soil as if waiting impatiently. He or she (males and females are indistinguishable) flew to my palm and politely took the first seed, then another. I was on a high for the rest of the day and called everyone I knew from Florida to Utah to share my good fortune.

Acorns are the scrub jays' primary food, but they also eat insects, berries, seeds, spiders, lizards, small snakes, and frogs. Each year they bury thousands of acorns from the oaks to maintain a stash. Many of the seeds aren't recovered and eventually sprout into new trees. Is this a cooperative system between the jays and the trees?

Family units of scrub jays have a strong social structure with three to eight birds per group. The center of the family is a breeding pair who stays together for life. Their offspring stay in the family territory and work as "helpers" for at least one year. Helpers contribute to the feeding and guarding of young scrub jays and help defend the family's territory.

Another typical scrub bird is the **rufous-sided towhee** (*Pipilo erythropthalmus*) which is now officially called the eastern towhee. In addition to their call (which sounds like, *drink your tea, drink your tea*), you can hear towhees rustling through dry leaves in search of insects.

Ground doves (*Columbina passerina*) are also commonly seen on the ground in scrub habitats.

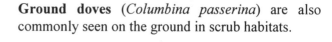

"Would you really like to live in a place where there is no mystery left?"
- M. Scott Peck, *On Heaven as on Earth*

Hot Life

Fire is an important part of scrub life. If scrub growth isn't interrupted by fire, the habitat grows tall and thick which is unsuitable for many of the scrub creatures, particularly the scrub jays. With thicker and higher plant growth, scrub jays' main competitors, the common blue jays, dominate. These blue jays become aggressive predators, and the scrub jay can't hide or survive in the unfamiliar tall scrub.

Saw palmettos (*Serenoa repens*) are one of the first plants to send up new growth after a fire. Their well-insulated centers, surrounded by a thick, waxy, moist coating are evidence of their survival skills.

The saw palmettos' intricate and strong root system stabilizes the sandy soil. Herb Allen says, "The roots of the palmetto hold Florida together."

The saw palmetto has an edible bud like the cabbage palm. But unlike the cabbage palm, saw palmettos can survive losing its terminal bud by sprouting new growth.

Borrowing Burrows

Dirt mounds in scrub are created by so many different animals that I can't easily figure out who the owners are. And to add to the complicated scheme of tunnels in this hot, sandy soil, gopher tortoise burrows are also used by snakes, lizards, frogs, spiders, possums, insects and mice. And there's more: even ghost crabs and armadillos add their tunneling to the scrub.

These community burrows protect the animals from heat and cold, as well as from fire and predators. All the digging helps return nutrients to the soil by recycling and aerating the sand. So, what appears to be a chaotic mess of mounds, tangles and leaf litter is really a creatively organized system of survival. Like my office.

Gopher tortoises' (*Gopherus polyphemus*) burrows are up to 35 feet long and more than a yard deep with "bedrooms" at the end. They dig down to moist, cool soil where they build their living quarters.

Adult gopher tortoises are tan-brown to gray-black and can grow to about 12 inches. Their front feet are flat like spades with hard nails, good for digging deep burrows. These tortoises communicate with each other by head bobs, touches and noises. Courtship signals include nipping, head bobbing and chin rubbing.

"Keep a quiet heart, sit calmly like a tortoise, walk sprightly like a pigeon, and sleep soundly like a dog."
- Li Chung Yun, Chinese herbalist

Community Spirit

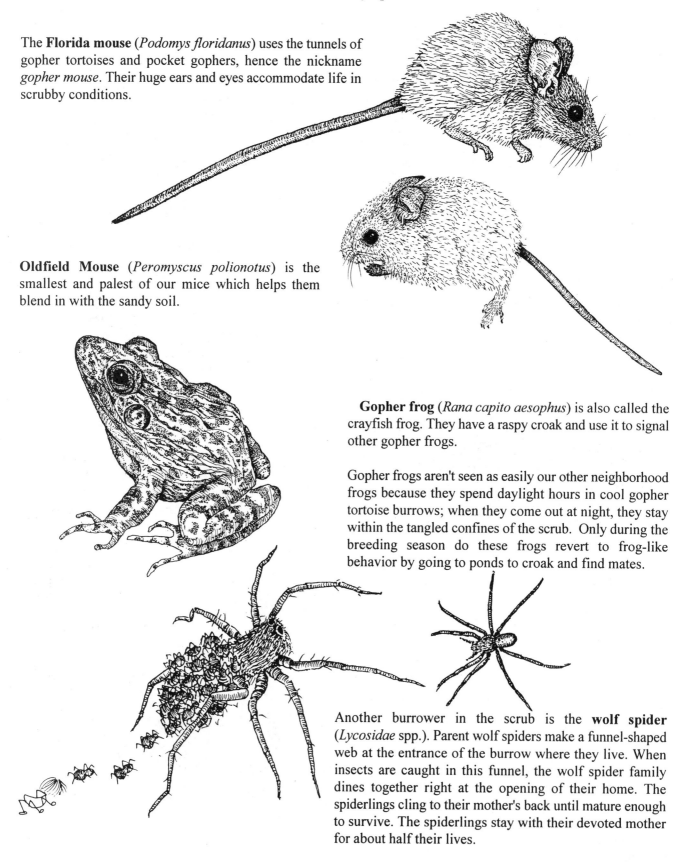

The **Florida mouse** (*Podomys floridanus*) uses the tunnels of gopher tortoises and pocket gophers, hence the nickname *gopher mouse*. Their huge ears and eyes accommodate life in scrubby conditions.

Oldfield Mouse (*Peromyscus polionotus*) is the smallest and palest of our mice which helps them blend in with the sandy soil.

Gopher frog (*Rana capito aesophus*) is also called the crayfish frog. They have a raspy croak and use it to signal other gopher frogs.

Gopher frogs aren't seen as easily our other neighborhood frogs because they spend daylight hours in cool gopher tortoise burrows; when they come out at night, they stay within the tangled confines of the scrub. Only during the breeding season do these frogs revert to frog-like behavior by going to ponds to croak and find mates.

Another burrower in the scrub is the **wolf spider** (*Lycosidae* spp.). Parent wolf spiders make a funnel-shaped web at the entrance of the burrow where they live. When insects are caught in this funnel, the wolf spider family dines together right at the opening of their home. The spiderlings cling to their mother's back until mature enough to survive. The spiderlings stay with their devoted mother for about half their lives.

"Birds are a good indicator of environmental problems, but amphibians are even better at the local level. Birds can fly away, but frogs must face the problem."
David Wake, Director of the Museum of Vertebrate Zoology at University of California, Berkeley

Scrub Neighbors

Pocket gophers (*Geomys pinetis*) are only about the size of my fist but they're capable of digging long tunnels running horizontally under the sand's surface, about a yard below. They have small eyes and ears, but long strong nails made for digging. They also have sharp teeth like beavers. Periodically they need to dig back to the surface to get rid of the excavated soil, leaving tell-tale signs of their burrowing: each return hole has a pile of dirt differently colored from the surface sand.

The skin around their cheeks is loose, so they can fill up with food to be stored in their burrows. Their name comes from their ability to turn their cheeks inside out to dump the food, like emptying pants pockets.

Nine-banded armadillos (*novemcinctus mexicanus*) can easily crawl through the prickly, tangled undergrowth of scrub areas, protected by their hard plates. (See page 54 for more on armadillos.)

One of the insects that armadillos look for at night is the **ground beetle** (*Calosoma scrutator*). It is a large (about an inch long), blackish-purple beetle with a green thorax. They hunt at night and hide under debris and leaves during the day.

The beetles' white eggs are as tiny as dust particles, and the larvae from the eggs are as thin as thread. When they grow to about an inch, they turn into pupae and then become adult beetles who return to the soil to hibernate. They come out at night to feed on soft larvae and caterpillars.

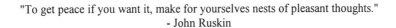

"To get peace if you want it, make for yourselves nests of pleasant thoughts."
- John Ruskin

Scrub Reptiles

Florida scrub lizards (*Sceloporus woodi*) (shown to the right) live in the hot sandy conditions of the scrub.

These lizards don't run after their prey, but rather, they sit and wait for insects like ants to appear and then they dart at their meal.

This pale brown, rough-scaled lizard looks like our more familiar fence lizard, but the scrub lizard has a prominent, dark brown line running along its side from head to tail. Male scrub lizards have a long blue area on each side of the belly and under their chin.

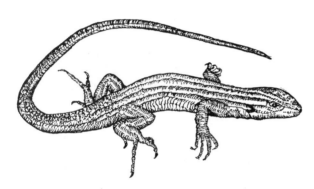

↑ Although not restricted to scrub, the **six-lined racerunner** *(Cnemidophus sexlineatus sexlineatus)* likes the hot, dry, sandy soil of scrub. They can survive higher temperatures than any other lizard in Florida. You can see them during the hottest time of the day dashing nervously through dead leaves, looking for insects. At night they stay under rocks and logs, or they stay in abandoned burrows.

Swimming in Sand?

Sand skinks (*Neoseps reynoldsi*) (shown below) only live in Florida. Their pale bodies blend well in the scrub's sandy soil. They have unusually tiny legs with only one toe on each front foot and two toes on the hind feet. Their eyes are small with transparent eyelids that work as goggles for swimming through the sugar sand during the heat of the day. After sunset, sand skinks come to the surface to prowl through leaf litter, looking for food such as termites and insect larvae.

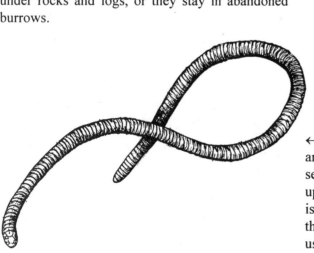

Heads or tails?

← **Worm lizards** (*Rhineura floridana*) are pale pink and are often mistaken for earth worms because of their segmented bodies. Which end is the head? Hard to tell, even up close, since the head has no eyes or ears, and its mouth is hidden underneath. If you look closely, you can see that the head is flatter than its tail, almost shovel-shaped. They use their snout like a spade to dig themselves underground.

"There are voices which we hear in solitude, but they grow faint and inaudible as we enter into the world.
- Ralph Waldo Emerson

More Scrub Reptiles

Legless lizards?

Eastern glass lizards' (*Phisaurus ventralis*) beautiful colors and designs attracted my attention when I first saw one crawling out from the thick scrub area near my house. It had no legs, so I thought it was a small snake. But when I looked closer, I noticed it had eye lids and ear openings, which were clues that it wasn't a snake. When I asked about it later, I found out that it is a common lizard in Florida.

Not limited to scrub habitat, glass lizards can also live in dunes, vacant lots and at the beach. Their name comes from their ability to easily break off their tails. This ability allows them to escape from attacks by diverting the predators' attention to the wiggling tail. By the time the attacker discovers it has caught a tail, the glass lizard has had time to hide. It will grow a new tail.

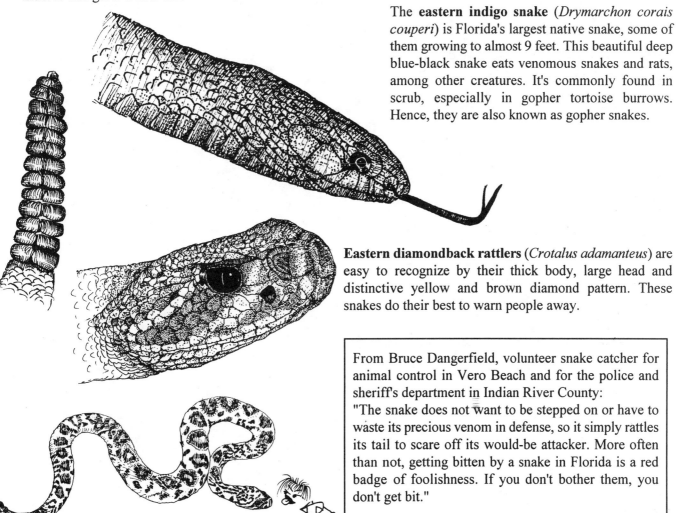

The **eastern indigo snake** (*Drymarchon corais couperi*) is Florida's largest native snake, some of them growing to almost 9 feet. This beautiful deep blue-black snake eats venomous snakes and rats, among other creatures. It's commonly found in scrub, especially in gopher tortoise burrows. Hence, they are also known as gopher snakes.

Eastern diamondback rattlers (*Crotalus adamanteus*) are easy to recognize by their thick body, large head and distinctive yellow and brown diamond pattern. These snakes do their best to warn people away.

From Bruce Dangerfield, volunteer snake catcher for animal control in Vero Beach and for the police and sheriff's department in Indian River County:
"The snake does not want to be stepped on or have to waste its precious venom in defense, so it simply rattles its tail to scare off its would-be attacker. More often than not, getting bitten by a snake in Florida is a red badge of foolishness. If you don't bother them, you don't get bit."

Like all pit vipers, rattlers use heat-sensitive pits in front of their eyes to sense danger or prey. These organs are so sensitive, a rattlesnake can be blindfolded and still be able to accurately strike its warm-blooded victim.

"Tell me what your vision of the future is, and I will tell you what you are."
-Theodosius Dobzhansky, *Mankind Evolving*

Scrub Plants

Sand live oaks (*Quercus geminata*) grow easily in sandy soil. They have the same artistic twists and turns as the **live oak** (*Quercus virginiana*) but in miniature. This is one of the many oaks that provides a good sanctuary for wildlife (even in backyards); its tangled branches form a thick haven of hiding places and food. Low plants, such as wild coffee, thrive under these oaks, and air plants use live oaks as host trees.

Orchard orioles, painted buntings and lots of other small birds are attracted to the oaks' acorns, not to eat the nut but to get the small worms inside. Humans can eat the oak acorns too. Pick a bunch and put them in a bowl of water. Throw away the ones that float. (They probably don't have any meat on them.) Keep the ones that sink and put them in a warm oven (about 150° to 200°) for 15 minutes. Crack and eat. (You might want to check for worms before you swallow.)

Prickly pear cactus (*Puntia compressa*) is shown to the far left.

Gopher apple (*Licania michauxii*) (shown above gopher tortoise's head) grows low enough for gopher tortoises and other small animals to get the fruit. This thick ground cover grows easily in sandy soil and is salt and cold tolerant.

"Carefully observe what way your heart draws you, and then choose that way with all your strength."
- Hasidic saying

More Scrub Plants

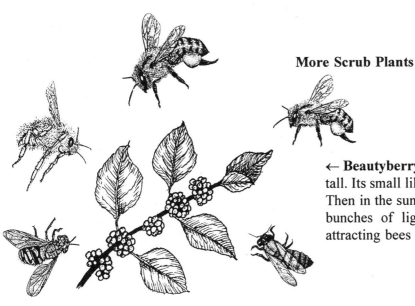

← **Beautyberry** (*callicarpa americana*) grows to about 6 feet tall. Its small lilac flowers blossom in the spring, luring birds. Then in the summer and fall (sometimes through the winter), bunches of light purple berries cluster around the stems attracting bees and other good pollinators.

← **Blazing star** (*Liatris tenuifolia*) is one of the more colorful blooming plants found in scrub habitat. Its rich lavender flowers attract butterflies such as gulf fritillaries.

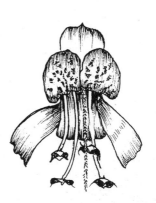

Scrub mint ↑ (*Dicerandra immaculata*) contains a chemical that repels insects.

Lichens (*Cladonia leporina, C. prostrata, Clandina evansii*) are a combination of fungi and algae. Bobbie Harrington says they're called lichens (pronounced *like-ins*) because the algae and fungi *like* being in a close relationship.

Lichens (shown right) grow on tree trunks, rocks, fences, and old logs. Their activities help the growth process for other plants. To survive, lichens only need light, air, and a few minerals from air and rain. Because they easily absorb whatever is in the air, including toxins, they are a sensitive index of air pollution.

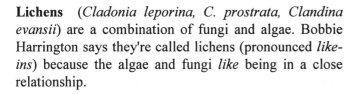

"If one advances confidently in the direction of his dreams, and endeavors to live the life he has imagined, he will meet with a success unexpected in common hours."
- Henry David Thoreau

NIGHTS

The live oak tree that grows just south of Mickey Collins' house in Melbourne Beach is draped with beautiful Spanish moss. When Mickey told me he wanted to yank off "all that ugly stringy stuff" because it was killing the live oak, I pleaded with him, "No, no, no. It doesn't hurt the tree." He didn't believe me, but luckily for our neighborhood, the Spanish moss is still there. So is the live oak, healthier than ever.

Spanish Moss

"The detractors of Spanish moss find it hard to give up the thought that the plant is a parasite up there sucking the lifeblood out of the trees it hangs on. That is not true."   Archie Carr in *A Naturalist in Florida*

Spanish moss (*Tillandsia usneoides*) is not a moss, nor is it Spanish. Rather, it's in the same family as air plants and pineapples (epiphytes). Notice the similarity to pineapple tops and air plants.

Air plants use other plants for support but not for nourishment. Spanish moss hangs from oaks such as **live oaks** (*Quercus virginiana*) and gets its food from the air and rain without damaging its host tree. It draws minerals from the dead cells and flaking bark that fall from higher branches of the tree and creates its own energy from the sun.

Minute scales cover the strands to absorb water and nutrients needed for the moss to survive, and tiny flowers grow from the end of the small side branches.

In early spring, these small green-yellow flowers sprout and give a wonderful, subtle, spicy scent. No other aroma affects me the way Spanish moss does. When I take my late night walks, I get lost inside my head and become unconscious of my surroundings. But when Spanish moss is blooming, its aroma draws my attention, reminding me to get out of my self-absorption.

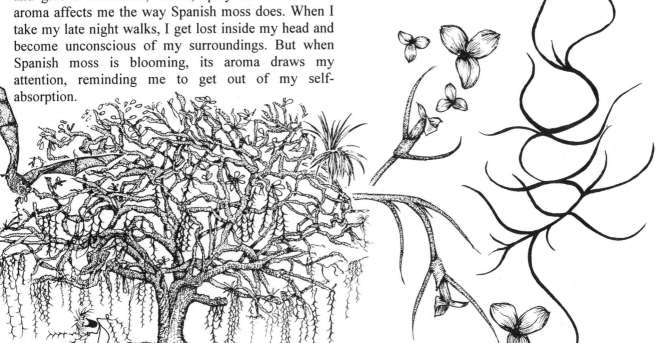

Spanish moss produces microscopic seeds that sprout wherever they scatter. Whenever I walk under the Spanish moss by Mickey's house, I position myself so I can feel it brushing my shoulder and hair. I like to think that I am a seed carrier and pollinator too, like the bats and birds. I like to encourage the moss to grow on other trees just to spite Mickey who still believes that Spanish moss is harmful.

For decades in early Florida, moss picking was a big industry with at least 10 moss factories. Moss trucks were commonly seen transporting moss. The moss was used for seat cushions because it was soft and springy.

"Spanish moss hangs deftly draped around the upper, middle and lower branches of trees as if master decorators had positioned it in place."
-Franklin Russel in *The Okefenokee Swamp*

Lost in Moss: Bats

Birds, reptiles and mammals depend on Spanish moss for nesting, shelter and food. Squirrels use the strings to tie their leaves and palm fibers together to build nests. Rat snakes climb through Spanish moss in search of bats. Ospreys and several other Florida birds use the strands to line their nests. Bats nest in the thick Spanish moss and bear their young in it. To study bats, bat researcher William Jennings paid moss factories to save the hidden bats for his research: 296 bats were found in one shipment.

One of the bats that uses moss to roost in is the **evening bat** (*Nycticeius humeralis*), one of the most abundant in Florida. These small, brown bats also live inside dead palm fronds.

Our bat populations are diminishing because of unfounded fears (bats were routinely killed because of superstitions and bad reputations), but we're finally realizing how many of our old ideas about bats are wrong and how beneficial they actually are to our neighborhoods.

At dusk in my neighborhood, I see bats darting frantically and erratically (or so it seems) in the sky. What are they doing? They snatch mosquitoes on the fly, zooming in on one and then backtracking to another faster than I can follow. Sometimes I see them flutter around street lights in search of bugs attracted to the light. They fly with their mouths open to let the insects in. This gives them that strange, mean look we see in many bat photographs.

Gator fans at the University of Florida stadium in Gainesville are no longer bothered by mosquitoes. The school built a 20-foot by 20-foot bat house which now houses 20,000 bats. The bats eat as many as *20 million insects every night*. Entomologist Ken Glover, the pest control manager at U of F said they no longer have to spray. Instead they rely on the bats for pest control.

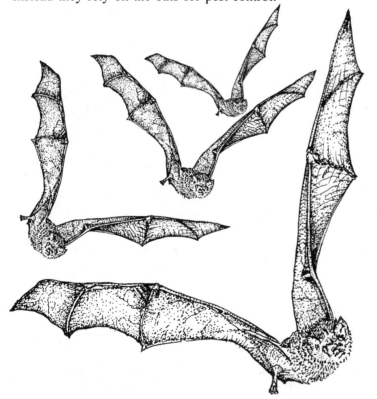

Another common bat in Florida is the **big brown bat** (*Eptesicus fuscus*). They live in buildings, wall spaces, and tree hollows. They mate in fall and winter. The females store the sperm until spring when they become active and pregnant. Females give birth to one or two babies in early summer.

During hibernation, bats' heart rate drops from 180 beats per minute to 3 beats per minute, and their respiration drops from *8 breaths per second to 8 per minute*.

Hibernating bats should never be awakened before they're ready: they use up their stored fat to revive themselves from this deep sleep and if they don't replace it right away, they die.

"Bats are feared only to the extent that they are misunderstood."
- Dr. Merlin Tuttle, bat expert and founder of Bats Conservation International and author of *America's Neighborhood Bats*

More Bats

Shown below is the **red bat** (*Lasiurus borealis*). Florida is also home to Seminole bats, hoary bats, eastern pipistrelles, Mexican free-tailed bats and yellow bats, among others.

Let your fingers do the flying. Bats actually use their fingers to spread their wings to fly. These fingers are similar to ours except they're longer and have a membrane connecting them. They're not fast flyers but they can maneuver better than hummingbirds.

Bats navigate by using **echolocation**. In an experiment in a dark room with *28 hair-thin wires* strung at random, and with *70 booming loudspeakers*, bats were able to fly through the maze of wires.

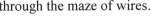

Rabies? "Less than half of 1 percent of bats contract rabies, a frequency no higher than that seen in many other animals. Like other mammals, they die quickly, but unlike even dogs and cats, rabid bats seldom become aggressive."
- Dr. Merlin Tuttle in *America's Neighborhood Bats*

Blind as a bat? Bats have excellent vision. In addition to their excellent eyesight, bats can also find their way through darkness by using a sophisticated echolocation system. They can navigate through dark skies and dodge wires and other obstacles with amazing accuracy.

Hawks, such as **red-tailed hawks** (*Buteo jamaicensis*) are on the lookout for bats just before dark and just after dawn. These hawks don't hunt the bats at night, but other predators such as screech owls take over, waiting by bat roosts for them to emerge.

"One doesn't recognize in one's life the really important moments -- not until it's too late."
-Agatha Christie

Night Birds

Eastern screech owls (*Otus asio floridanus*) are only about 7 to 10 inches tall so they're not as noticeable as the great horned owl. The adults' call is a quavering trill like a horse whinny. Only baby screech owls screech. A screech owl chick is shown to the right.

Screech owls (adult shown left) are fairly common in my neighborhood but I didn't know that for the first few years I lived in Melbourne Beach. I couldn't identify the eerie wail I heard at night every so often. It sounded like the wailing of a human in utter agony. When I discovered this piercing sound came from a screech owl, I was surprised, expecting a bird much larger than the 7-inch pipsqueak in my yard.

Screech owls nest in tree holes carved out by woodpeckers or in natural cavities.

Great horned owls (*Bubo virginianus*) (shown left) are the largest of all our Florida owls, growing to about 2 feet. Their "horns" are really tufts of feathers. Their ears are farther down on their head. When they raise their tufts, they are displaying their aggressive behavior. The tufts are also used to gather and funnel sounds down to their keen ears.

The eyes of a great horned owl are so large they leave no room for muscles to move the eyes. Instead, the owl uses 14 vertebrae in its neck (humans only have 7) to turn its head in the characteristic owl swivel.

These sharp eyes and ears make them excellent hunters, so they can hunt for a variety of food -- skunks, rabbits, rodents and birds.

Female great horned owls will lay their eggs in unused hawk, crow, or heron nests. Jim Angy put an old laundry basket stuffed with leaves and twigs in a tree near owl territory. Every year since then, a pair of great horned owls has used the basket to raise their chicks.

"We could never learn to be brave and patient if there were only joy in the world." - Helen Keller

Night Plants and Insects

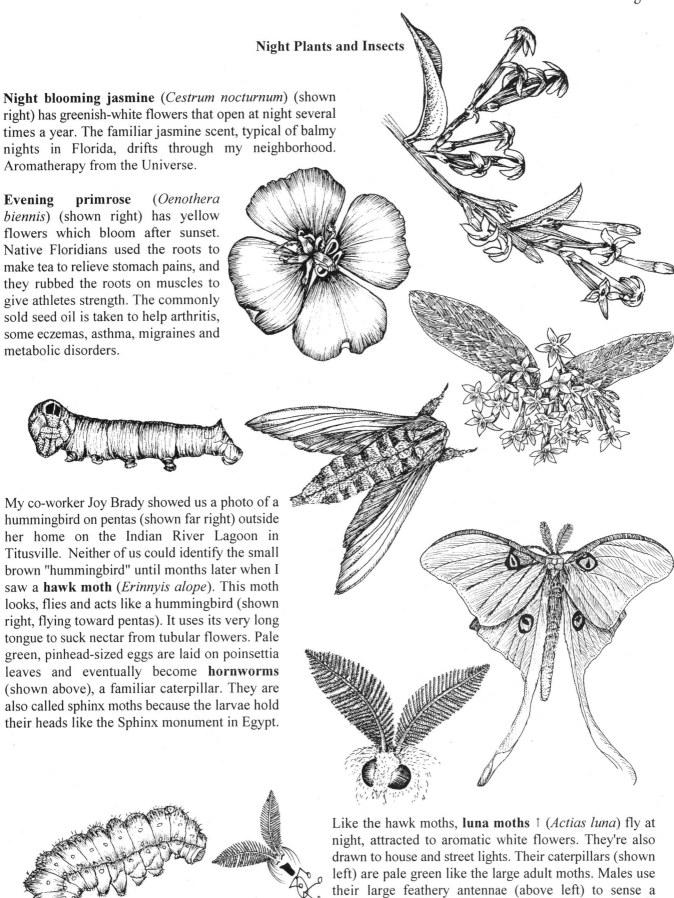

Night blooming jasmine (*Cestrum nocturnum*) (shown right) has greenish-white flowers that open at night several times a year. The familiar jasmine scent, typical of balmy nights in Florida, drifts through my neighborhood. Aromatherapy from the Universe.

Evening primrose (*Oenothera biennis*) (shown right) has yellow flowers which bloom after sunset. Native Floridians used the roots to make tea to relieve stomach pains, and they rubbed the roots on muscles to give athletes strength. The commonly sold seed oil is taken to help arthritis, some eczemas, asthma, migraines and metabolic disorders.

My co-worker Joy Brady showed us a photo of a hummingbird on pentas (shown far right) outside her home on the Indian River Lagoon in Titusville. Neither of us could identify the small brown "hummingbird" until months later when I saw a **hawk moth** (*Erinnyis alope*). This moth looks, flies and acts like a hummingbird (shown right, flying toward pentas). It uses its very long tongue to suck nectar from tubular flowers. Pale green, pinhead-sized eggs are laid on poinsettia leaves and eventually become **hornworms** (shown above), a familiar caterpillar. They are also called sphinx moths because the larvae hold their heads like the Sphinx monument in Egypt.

Like the hawk moths, **luna moths** ↑ (*Actias luna*) fly at night, attracted to aromatic white flowers. They're also drawn to house and street lights. Their caterpillars (shown left) are pale green like the large adult moths. Males use their large feathery antennae (above left) to sense a female's scent.

"Smells are surer than sounds and sights to make heart strings crack." -Rudyard Kipling

Night Mammals

Raccoons (*Procyon lotor eleucus*) don't need a lot of introduction. They are the most urbanized wild creature in Florida's neighborhoods. Around our neighborhood, they get into attics, garages, birdfeeders, my cat's food bowl and especially our garbage cans.

Nine-banded armadillos' (*novemcinctus mexicanus*) name comes from the Spanish word for "little armored thing." The overlapping bands of their armor allow them to bend and maneuver through tricky areas. Their bands are actually made of 2500 overlapping plates. Their armor isn't hard enough to protect them from predators' bites, but it gives them the advantage of being able to crawl through prickly, thick tangles of scrub to escape danger.

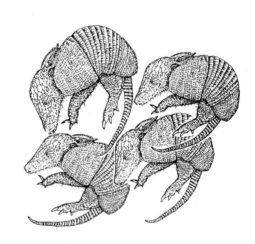

Armadillos are good diggers and spend their days inside burrows dug by their strong, sharp claws. They usually come out at night to dig for beetles, grubs, ant eggs, and sometimes roots and berries. They are wonderful insect controllers (*particularly fire ants*) since they can eat 200 pounds of insects a year.

Females always give birth to four identical babies of the same sex. This unusual pattern results from one egg (per year) dividing after fertilization, eventually producing four embryos. Another unusual reproductive trait is their irregularity in pregnancy -- embryo development can be delayed by two years after mating.

The **possum** (*Didelphis marsupialis pigra*) is the only marsupial that lives in North America. Native Americans named them *apasum* meaning white animal. Their favorite foods are insects like beetles, butterflies and grasshoppers, but they also like worms, lizards, frogs and small snakes. Often when they're dining on road kill, they become road kill themselves.

Female possums give birth to as many as 20 young at a time, only 13 days after mating. Each baby is the size of a bumblebee, and it crawls into the mother's pouch where 13 nipples provide milk for about 2 months. If the mother gives birth to more than 13 babies, the rest don't survive since each baby who gets a nipple latches on for the duration.

Playing Possum?

When a possum is threatened, it poses as dead. It goes into a catatonic, limp heap with mouth open, teeth exposed, tongue hanging out, and sometimes even a greenish ooze comes from an anal gland. But scientists now think that possums aren't clever enough to devise such deceit, but rather, they faint from fright.

"Atoms from a dead possum may become part of a berry bush. The processing of life after death is so hidden to us that we rarely think about it, and sometimes we don't want to." - Valerie Harms in *The Almanac of the Environment, The Ecology of Everyday Life*.

Greens and Cubans: Frogs and Anoles (pronounced ah-NO-lays)

Green anoles (*Anolis carolinensis*) (shown right) are the most common lizards in my neighborhood. I used to think they were chameleons because they could change color so fast -- I see them change from green to brown with white patterns, back to green, all within minutes.

Anoles have adhesive pads on the bottom of their toes, allowing them to run, dodge and turn quickly even on slick surfaces. The males have a throat fan called a *dewlap* (shown above left) which is pink or orange. They fan out this dewlap to communicate; either to warn trespassers to stay away or to send a message to females.

Why do they sit motionless for hours? Instead of actively searching for food, they hang around where insects gather, waiting for the food to come to them.

Anoles shed their skin and it falls away from their body in little pieces. Some of these pieces look like the plastic circles that fall from the bottom of plastic shopping bags. Sometimes I see anoles trying to eat those plastic patches, mistaking them for their skin which it eats for the nutrients. (They can choke on the plastic, so it's a good deed to make sure these pieces don't lie around.)

The brown **Cuban brown anole** (*Anolis sagrei sagrei*), (shown above) is new to Florida and more aggressive than the green anole. Some people think the Cuban brown is taking over the green's territory, because these newcomers breed faster than the greens and can overrun them. Cuban anoles have a distinctive brown and white design on their backs, but unlike the green anoles, they cannot change to green. Sometimes the males have a ridge that looks like loose, pinched skin sticking up along the back.

Lizard poop looks like Rice Krispies® dipped in dark chocolate. (Each dropping is tipped in white.) If you have this evidence in your home, it's good sign that lizards live with you -- and they're probably eating insects you'd rather not have around. (Cockroach poop looks like All Bran® that has been sitting in milk for awhile.)

The **Cuban tree frog** (*Osteopilus septentrionalis*) has well developed toe pads typical of Florida's tree frogs. These disk-like pads secrete mucus that helps the frog stick to almost any surface, as we can see from the way they cling to windows and glass sliding doors. They can also change colors rapidly from pale tan to green to dark brown. If not visible all the time, they're definitely heard -- their loud and demanding croaks carry easily through the night air, especially after rains.

This 2- to 3-inch **green tree frog** (*Hyla cinerea*) is one of the most familiar frogs to Floridians. They range from different greens to brown, but they all have the distinctive white stripe running along their sides from front to back.

"I have always liked frogs. I liked them before I ever took up zoology as a profession; and nothing I have ever had to learn about them since has marred the attachment. I like the looks of frogs, and their outlook, and especially the way they get together in wet places on warm nights and sing about sex. The music frogs make at night is a pleasant thing, full of optimism and inner meaning..."

-Archie Carr in *The Windward Road*

Cockroaches: Contemptible Little Caterpillars

Cockroaches are survivors; fossil remains from 400 million years ago have been found. Florida has 41 species of cockroaches, more than any other state. Twenty-seven are native to Florida and fourteen are immigrants. Most can survive without food for a few months and without water for about 30 days. They eat just about *anything*. The Spanish word for cockroach, *cucaracha,* means *contemptible little caterpillar.*

"Cockroaches have changed little over the aeons. Like the designers of Volkswagen Beetles, they have stuck with what appears to be a winning formula, to which they've added countless refinements over time. The result of all this fine-tuning? An inconspicuous but well-equipped creature with alien features and abilities far beyond our own."   David George Gordon in *The Compleat Cockroach*

Cockroaches have sensitive hearing and can detect tiny movements of air by using 220 microscopic hairs on each of two rear appendages.

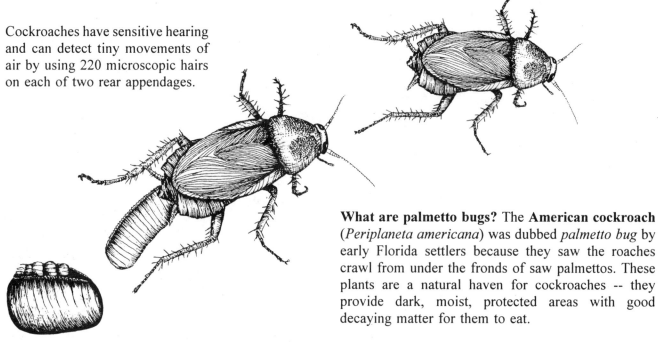

What are palmetto bugs? The **American cockroach** (*Periplaneta americana*) was dubbed *palmetto bug* by early Florida settlers because they saw the roaches crawl from under the fronds of saw palmettos. These plants are a natural haven for cockroaches -- they provide dark, moist, protected areas with good decaying matter for them to eat.

Shown above is a cockroach egg case, *ootheca* (oh-a-theé-kah) in which two rows of eggs are laid. Each case is about ⅓ inch and looks like a shiny, dark brown purse. The female lays 10 to 30 eggs in each case. Some females carry the ootheca around (as shown above) until the roach babies are born. All newborns are wingless. Nymphs take about 12 months to reach maturity.

Robo-roach? Palmetto bugs are used in Japan for research. Electronic backpacks are surgically implanted to control the roaches' movements. Why? Roaches can crawl through tiny spaces to search for earthquake, kidnapping, or hostage victims. Spies would also have a new means to break into espionage cracks.

"Recent research has established that a cockroach breaks wind every 15 minutes. It also continues to release methane for 18 hours after death. Insect flatulence is said to account for 20% of all methane emissions on earth, placing termites and cockroaches among the biggest contributors to global warming."   - David George Gordon in *The Compleat Cockroach*

More Cockroach Stuff

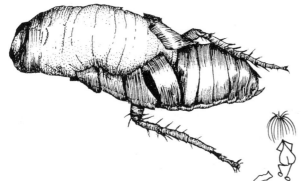

Cockroaches come out at night after hiding in dark, damp places all day. Shown to the left is a cockroach nymph molting. They go through several molting phases as they grow, coming out tender and pale from their old skin; within a few hours they change to brown. Whenever I get uncomfortable in my old skin and have to leave the old stuff behind, I feel a little tender too -- until I get used to the new me and then I wonder what took me so long to crawl out of the old. Cockroaches eat their old skin for the nutrition.

Oriental cockroaches (*Blatta orientalis*) are between 1 and 2 inches, black to reddish brown, and glossy. They like to hang around wet areas in kitchens and bathrooms. Females carry their egg case with them for a few days. But then she tucks it away in a corner and abandons it forever. Each egg case of this roach contains about 10 eggs.

German cockroaches (*Blatta germanica*) (shown right) are also small (usually a little more than half an inch) and will also eat anything available. They can walk on slick vertical surfaces because they have special adhesive pads on their legs.

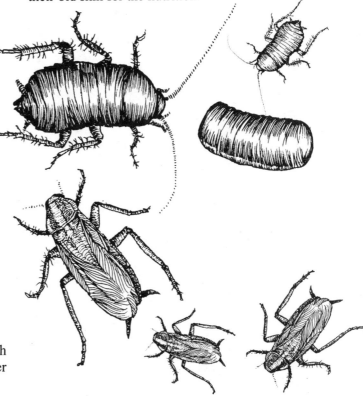

The smallest cockroach is the **brown-banded** cockroach (*Suppela longipalpa*) (shown above). Unlike the other roaches, they prefer warm, dry places.

Cockroach control? House geckos love to eat roaches and can squirm into tight hiding places. Some Floridians have been buying geckos as a way to control roaches. Geckos only come out at night to hunt, so they're not nearly as visible as the anoles.

I never bought any geckos, but I find their eggs in corners and inside drawers in a couple rooms of my house. Females lay two white eggs, each about ⅓ inch. Sometimes I see a pregnant gecko on my window, and I can see her eggs bulging through her skin, as shown below. This species has expanded faster than other geckos in Florida, because they are all self-fertilizing females. One individual can start a new colony.

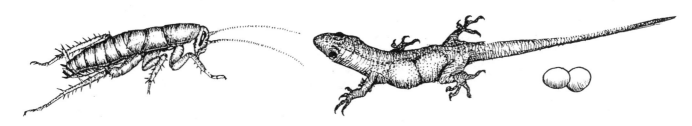

"Ask yourself how you feel about cockroaches. Do you shiver at the mere mention of their name? Could you kill one with your bare hand? Would you rather not be talking about this?" - David George Gordon in *The Compleat Cockroach*

Busy Night Life

Silverfish (*Lepisma saccharina*) come out at night to feed on just about anything: pet food, vegetables, books, cereal, paste. They have long thready antennae to feel their way in the dark. They're good at hiding in tiny spaces, but we find them in sinks or bathtubs after they fall in and can't get out.

Silverfish drop little white eggs one at a time, here and there, week after week, so they're hard to control. Jaye Wright in *Florida TODAY* Newspaper recommends dusting with the natural insecticide pyrethrin powder, which also helps control fleas.

Most of us are familiar with **no-see-ums** (*Culicoides furens*) (shown left) whose bite is far out of proportion to their size. They're so tiny, one of them could hide behind this period ↘. Unlike mosquitoes (who pierce our skin) no-see-ums cut with pincers, causing an intense hot pinch. Some people are sensitive to no-see-um bites and get welts.

Mosquitoes (*Aedes* spp.) used to be so thick in Florida that people would be covered with them after being outside for only a few minutes. Can you imagine?

Another insect that flies around street lights is the **lacewing** (*Chrysopa bicornea*), shown left. Lacewings eat lots of aphids, scales and mites.

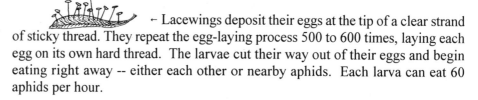

← Lacewings deposit their eggs at the tip of a clear strand of sticky thread. They repeat the egg-laying process 500 to 600 times, laying each egg on its own hard thread. The larvae cut their way out of their eggs and begin eating right away -- either each other or nearby aphids. Each larva can eat 60 aphids per hour.

How do lacewings avoid insect-eating bats? Lacewings are sensitive to the high-frequency sounds emitted by bats for echolocation. When a bat approaches, the lacewing hears it, and plunges to the ground with folded wings. It stays on the ground until the bat flies away.

Mole crickets (*Gryllotalpa hexadactla*) (shown below) stay in tunnels in the soil most of the time. But in the spring if I see pieces of mulch move, I check it out and usually discover a wonderfully bizarre mole cricket. When I look closely at their amazing bodies, I see the creativity that went into the making of this creature: strong claws like a mole, smooth outer shell like a horseshoe crab and a husky abdomen like an ox.

"And I took the road less traveled, and that made all the difference."   - Robert Frost

Scary but Harmless

Slugs aren't insects, but rather descendants of land snails (which descended from sea snails). They are mollusks, like clams, scallops, and whelks, except they don't have a shell to protect their soft bodies. Those soft vulnerable bodies would dry out quickly in Florida's hot sun. That's why we only see them at night.

Droolers

In Spanish, slugs are called *babosas* from the verb meaning *to drool*. When I walk at night, I see wavy, thin strands of shiny trails on the sidewalk. These long trails are made from their slime. Slugs secrete lots of mucus which helps them move from place to place. The slime is secreted from a gland under the slug's head, and it flows down to the slug's muscular foot.

Slugs' mucus also protects them from diseases. Scientists are discovering that slug slime may help mucous-related diseases in humans such as cystic fibrosis.

Mucus also protects their soft vulnerable bodies from cuts and bruises. Gayle Mercurio and her biology class at Johnson Jr. High School watched a slug crawl over the edge of a new razor blade without scratching its flesh, protected by its own slime.
Slugs' eyes are at the tips of their tentacles and, in some species, also function as infrared receptors so they can hunt in the dark.

As interesting as they are, slugs aren't usually welcome in most yards. They destroy plants by eating tender leaves. But they only feed at night, so to catch them, you have to follow their shiny trail with a flashlight. Bill Zak in *A Field Guide to Florida Critters* recommends catching them by putting out a dish of beer since snails and slugs are attracted to beer. It's actually the yeast in the beer that attracts the slug, so you could also put out a package of yeast dissolved in sugar water.

Slug eggs above.

Whip-Scorpions (*Mastigoproctus giganteus*) (shown left) are also known as **vinegaroons** because they emit a strong acid smell like vinegar when they're disturbed.

Vinegaroons are nice to have around -- they eat insects we'd rather not see such as cockroaches. Their pincers look mean, but they're harmless and contain no poison. Jennifer Curry saw a whip scorpion more than 6 inches long!

Earwigs (*Lapidura riparia*) (shown right) are one of Florida's most common insects. They hide during the day but come out at night to scavenge. Their rear end pincers can grab onto things but aren't strong enough to even pinch or hurt a finger.

Springtails (*Isotoma* spp.) (Shown left) are easy to overlook because they are so small: a big one is about the size of a rye seed. Sometimes I think the dirt is moving when lots of them are active.

"Trifles make perfection, though perfection is no trifle." - Michelangelo

Three Widows

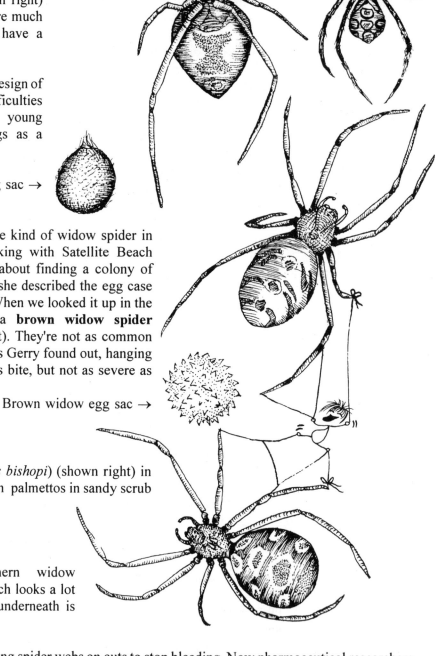

You won't see marks or swelling from the bite of a **black widow spider** (*Latrodectus mactans*), but you'll probably get severe abdominal pains as if you were having an appendicitis attack. Muscle pain all the way to the soles of your feet accompanied by dry mouth and sweating are other symptoms. (The good news is that recovery only takes a few days.) Females (shown right) are the biters. Males (shown far right) are much smaller than the ½-inch females and have a different pattern and different coloring.

Males lack the distinctive red hourglass design of the female. Sometimes you may have difficulties determining this spiders' sex, since young females often have the same markings as a mature male adult. Interesting.

Black widow egg sac →

I didn't know that we had more than one kind of widow spider in Florida until one day when I was talking with Satellite Beach Librarian Gerry Petrovic. She told me about finding a colony of black widows in her mailbox, but then she described the egg case as a little ball with spikes all around it. When we looked it up in the spider book, we discovered it was a **brown widow spider** (*Lactrodectus geometricus*) (shown right). They're not as common as the black widows, but they are seen, as Gerry found out, hanging around buildings. They have a poisonous bite, but not as severe as the black widows.

Brown widow egg sac →

We also have **red widows** (*Lactrodectus bishopi*) (shown right) in Florida. Who knew? But these only live in palmettos in sandy scrub areas.

Northern Florida hosts the northern widow (*Lactrodectus variolus*) (not shown) which looks a lot like our red widow except the pattern underneath is split into two or three smaller patches.

Florida Cracker folk wisdom advises putting spider webs on cuts to stop bleeding. Now pharmaceutical researchers at the University of Wyoming have found evidence that doing this is a good thing: spider silk effectively resists bacteria and viruses. The researchers also discovered that spider silk can be used as sutures in surgery.

"We become what we think about. If we don't think at all, we don't become anything at all."
- *Earl Nightengale's Greatest Discovery*

Stingers

The sting of a **scorpion** (*Centruroides gracilis*) comes from the gland at the end of its tail. The sting is painful but not dangerous. The worst that can happen is you'll see some swelling around the sting.

I like having scorpions around the house because they catch insects and spiders with their pincers. The scorpions I see are usually about an inch long.

Scorpions are nocturnal and stay well hidden. They can detect our slightest movements by using tiny structures called *pectines* (shown right): a pair of combs that touch the ground as the scorpion walks, sensitive enough to detect vibrations of their prey.

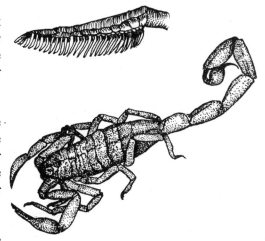

Scorpions have a long courtship. The male and female circle around each other with their tails up. The male grabs the female with his pincers, and they walk backward and forward together for hours. Eventually the male drops a sperm package on the ground and maneuvers her until she can take the sperm into her body.

When the eggs hatch, the tiny scorpions climb on top of the female's back and cling to her while maturing.

In 1993, Nick Sabin unknowingly met a **brown recluse** (*Loxosceles reclusa*) while he was working in his attic. He felt a hot jab in his back, but ignored it, thinking he had backed into one of the long nails sticking out in the crawl space. He continued to ignore the pain in his back muscle, even as it intensified. But when he went to bed, he couldn't sleep because he began to feel really sick as if he had the flu. He finally went to the doctor, who diagnosed Nick as having been bitten by a brown recluse; the doctor was able to make the accurate diagnosis from Nick's *missing flesh*. (Clueless Nick hadn't noticed the *two holes* in his back after he was bitten.)

So. Recluse spider bites are severe and dangerous. People die from them. Typically, these spiders aren't aggressive, but when threatened (either real or imagined), they will defend themselves by using their powerful venom.

Their bite causes a sharp sting, then surrounding pain. Within 24 to 36 hours, fever, chills, nausea, and joint pain will set in. Ask Nick about the pain. A small blister surrounded by a big swollen area is another indication that you've been bitten by a brown recluse. Healing takes 6 to 8 weeks, and the bite will leave a scar from the venom's killing effect.

Where to look for them? Like me, they like cluttered, messy areas with lots of little places to hide. To identify one, notice that their body is orange to dark brown. They have three pairs of eyes and a violin-shaped design on their head.

Actual size

"The way I see it, if you want the rainbow, you gotta put up with the rain." - Dolly Parton

Bibliography

Ajilvsgi, Geyata. *Butterfly Gardening for the South*. Dallas: Taylor Publishing Co., 1990.

Anderson, Robert. *Florida Snakes*. St. Petersburg: Great Outdoors Publishing Co., 1989.

Bell, C. Ritchie and Taylor, Bryan J. *Florida Wild Flowers*. Chapel Hill: Laurel Hill Press, 1982.

Beriault, John G. *Planning and Planting a Native Plant Yard*. Spring Hill: Florida Native Plant Society, 1988.

Carr, Anna. *Rodale's Color Handbook of Garden Insects*. Emmaus: Rodale Press, 1979

Comstock, John H. *The Spider Book*. Ithaca: Comstock Publishing Associates, 1948

Cox, Jim. *Birdwatching Basics*. Tallahassee: The Florida Game and Fresh Water Fish Commission Nongame Wildlife Program.

Dennis, John V. *A Complete Guide to Bird Feeding*. New York: Alfred A. Knopf, 1994.

Gingerich, Jerry. *Florida's Fabulous Mammals*. Tampa: World Publications, 1994.

Gordon, David G. *The Compleat Cockroach*. Berkeley: Ten Speed Press, 1996

Huegel, Craig. *Butterfly Gardening with Florida's Native Plants*. Orlando: Florida Native Plant Society, 1991.

Florida Plants for Wildlife. Orlando: Florida Native Plant Society, 1991.

Kale, Herbert W., and Maehr, David S. *Florida's Birds*. Sarasota: Pineapple Press, 1990.

MacCubbin, Tom. *Florida Home Grown 2: The Edible Landscape*. Orlando: Sentinel Books, 1989.

Maxwell, Lewis S. *Florida Insects*. Tampa: Lewis Maxwell, 1990.

Pringle, Lawrence. *Here Come the Killer Bees*. New York: William Morrow and Co., Inc., 1986.

Milne, L. and M. *The Audubon Society Field Guide to North American Insects and Spiders*. New York: Alfred A. Knopf, 1980.

Sattler, Helen R. *The Book of North American Owls*. New York: Clarion Books, 1995.

Stiling, Peter D. *Florida's Butterflies and Other Insects*. Sarasota: Pineapple Press: 1989.

Stokes, Donald W. *A Guide to the Behavior of Common Birds*. Boston: Little, Brown and Company, 1979.

Tuttle, Merlin. *America's Neighborhood Bats*. Austin: University of Texas Press, 1988.

Williams, Winston. *Florida's Fabulous Reptiles and Amphibians*. Tampa: World Publications, 1991.

Woolfenden, Glen. and Fitzpatrick, John W. *The Florida Scrub Jay*. Princeton: Princeton University Press, 1984

Zak, Bill. *A Field Guide to Florida Critters*. Texas: Taylor Publishing Co., 1986.

Index

Little Larry
She's back and she's as goofy as she wants to be.

The little figure who shows up on every page of this book is Little Larry. She's a creative spirit who knows no limits, just like her namesake, the late Larry McGrath. As in the previous books in this series, Little Larry does anything she wants to do, especially if it's ridiculous. She thinks nothing of borrowing butterfly wings, juggling slug eggs, imitating fire ants, crawling around with spiderlings or playing football with fly larvae.

The author and the publisher refuse to accept responsibility for any actions of this ½-inch stick figure.

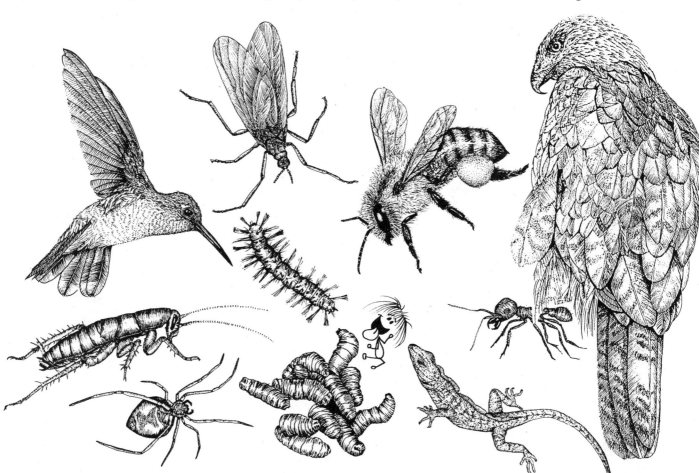

The photographs on the front and back are by Jim Angy, wildlife photographer and dedicated naturalist. On the front cover is a young raccoon and on the back is a scrub jay chick.

In addition to providing these photographs, Jim teaches me to see details clearly and objectively. I no longer say, "This is a *good* bug and this one is a *bad* bug."

And he pulls in my tendency to make grand sweeping statements like, "The giant swallowtail is the *largest* butterfly in the world." (He informs me that it is the *second largest butterfly in Florida*.)

Jim's accuracy in words and thoughts is reflected in his clear photographs.